Topics on Quantum Information Science

Edited by Sergio Curilef
and Angel Ricardo Plastino

Published in London, United Kingdom

IntechOpen

Supporting open minds since 2005

Topics on Quantum Information Science
http://dx.doi.org/10.5772/intechopen.94639
Edited by Sergio Curilef and Angel Ricardo Plastino

Contributors
Steven Duplij, Raimund Vogl, Gerard 't Hooft, Jesús Lacalle, Mrittunjoy Majumdar, Gopalan Raghavan, Amitabha Gupta, Oliver Keith Baker, Sergio Curilef, Angel Ricardo Plastino

First published in London, United Kingdom, 2021 by IntechOpen
IntechOpen is the global imprint of INTECHOPEN LIMITED, registered in England and Wales, registration number: 11086078, 5 Princes Gate Court, London, SW7 2QJ, United Kingdom
Printed in Croatia

British Library Cataloguing-in-Publication Data
A catalogue record for this book is available from the British Library

Additional hard and PDF copies can be obtained from orders@intechopen.com

Topics on Quantum Information Science
Edited by Sergio Curilef and Angel Ricardo Plastino
p. cm.
Print ISBN 978-1-83968-768-6
Online ISBN 978-1-83968-769-3
eBook (PDF) ISBN 978-1-83968-770-9

For EU product safety concerns:
IN TECH d.o.o., Prolaz Marije Krucifikse Kozulić 3, 51000 Rijeka, Croatia, info@intechopen.com or visit our website at intechopen.com.

Meet the editors

Sergio Curilef is a full professor in the Physics Department, Universidad Católica del Norte (UCN), Antofagasta, Chile. He obtained his Ph.D. in Physics from the Brazilian Center for Research in Physics (CBPF)-RJ, Brazil, in July 1997. His main research areas are statistical physics and quantum mechanics of systems with long-range interactions, particles in the magnetic field, and nonlinear diffusion. He has recently developed research on complex systems and applications to science and engineering. He is the principal investigator in several research projects, and he did a sabbatical year at the University of Pretoria, South Africa, in 2007. Dr. Curilef has also created undergraduate and postgraduate programs in physics at his university, and he is currently an academic counselor of the UCN Higher Council.

Angel Ricardo Plastino is a professor in the Departamento de Ciencias Básicas de la Universidad Nacional del Noroeste de la Provincia de Buenos Aires (UNNOBA), Argentina. He has a Ph.D. in Astronomy from the Universidad Nacional de La Plata, Argentina. His main research interests center on the multidisciplinary applications of ideas and techniques inspired by information theory. He has published more than 200 articles in peer-reviewed research journals on topics related to the physics of information, theory of quantum entanglement, foundations of quantum and statistical mechanics, physics of complex systems, and dynamical and mathematical aspects of astronomy.

Contents

Preface

Quantum information is one of the most active and rapidly advancing areas of physics. Developments in this field have a dual character. On one hand, exploring the information transmission and processing capabilities allowed by the basic laws of quantum physics may lead to new and revolutionary quantum information technologies. On the other hand, understanding the information-theoretical aspects of physical systems and processes contributes to a deeper understanding of physics itself. Due to these complementary facets, quantum information constitutes a rich and fertile field attracting a wide spectrum of researchers, ranging from engineering-motivated technologists interested in practical innovations to philosophically inclined theoreticians interested in foundational issues. The chapters in this book constitute a stimulating sample of the different aspects of quantum information science. Contributions by experts discuss a variety of topics, including investigations dealing with the implementation of quantum technologies and works exploring fundamental problems at the very frontiers of contemporary physics.

The introductory chapter emphasizes some features from a historical perspective on the physics of information and quantum mechanics. After this discussion, several additional chapters cover topics related to recent advances in the modeling and application of quantum information science.

In Chapter 2, Prof. ’t Hooft proposes a new theoretical explanation for quantum physics based on classical and deterministic models.

In Chapter 3, Prof. Majumdar formulates nested multilevel entanglement and discusses it in Matryoshka states.

In Chapter 4, Prof. Gupta presents some foundational issues in quantum information science, dividing his discussion into three parts.

In Chapter 5, Prof. Lacalle addresses the challenge of making quantum computing a reality, discusses the control of quantum errors, and presents a road map to quantum computing.

In Chapter 6, Profs. Duplij and Vogl propose a concept of quantum computing that incorporates a kind of uncertainty, the vagueness, introducing obscure qudits, which are simultaneously characterized by a quantum probability and a membership function.

In Chapter 7, Prof. Raghavan pays attention to a looming threat over current methods of data encryption through advances in quantum computation; due to this, physically assured privacy is provably secure only in theory and not in practice. The author includes a brief overview (not a review) of device independence and the conceptual and practical difficulties.

Finally, in Chapter 8, Prof. Baker demonstrates that several anomalies seen in data from high-energy physics experiments have their origin in quantum entanglement and quantum information science more generally.

Sergio Curilef
Departamento de Física,
Universidad Católica del Norte,
Antofagasta, Chile

Angel Ricardo Plastino
Universidad Nacional del Noroeste de la Provincia de Buenos Aires,
Junin, Argentina

Section 1

Introduction

Chapter 1

Introductory Chapter: Physics of Information and Quantum Mechanics - Some Remarks from a Historical Perspective

Sergio Curilef and Angel Ricardo Plastino

1. A bit of history

Ideas and techniques coming from information theory are nowadays gaining prominence in physics [1–6]. Quantum information, in particular, is one of the most active and rapidly advancing fields in the physical sciences [2, 6]. This state of affairs results from intricate historical developments that cannot be explained in a few pages. It is useful, however, in order to put quantum information science in a wider context, to try to summarize some of the main events that led to the present role of the concept of information in general physics and, in particular, in quantum physics. We shall briefly discuss and provide a short overview, within a historical perspective, of some basic features of the fields of the physics of information and quantum information. From the very inception of Shannon's information theory more than 70 years ago [7, 8], scientists found it intriguing that the engineering-motivated quantitative measure of information derived by Shannon is mathematically identical to the expression for entropy proposed several decades before by Boltzmann and by Gibbs. It is surprising that two conceptually different quantities, arising independently from completely different motivations in two unrelated fields, one belonging to pure science, and the other to engineering, share the same mathematical form. Various lines of thought developed in the subsequent years suggested that, rather than being just a superficial formal coincidence, the similarity indicates that there is a deep connection between information theory and physics. Physicists started paying serious attention to Shannon's information theory in the 1950s, thanks to a large extent to the pioneering efforts of Jaynes, who advanced a reformulation of statistical mechanics based on concepts from information theory [9, 10]. The basic idea behind Jaynes proposal is that the entropy associated with a macroscopic description of a physical system is actually a measure of the missing information about the system's precise microscopic state. Based on the connection between information and entropy Jaynes advanced the principle of maximum entropy (MaxEnt) as a guiding rule in statistical mechanics, for identifying the least biased statistical description of a physical system compatible with the available incomplete data. Later Jaynes promoted MaxEnt as a general principle of statistical inference. Some commentators do not include the works of Jaynes among the sources of the physics of information and of quantum information. Jaynes' works, however, were essential in propagating the notion that information theory is important for understanding fundamental aspects of physics. Ideas revolving around Jaynes' information-theoretical approach to statistical mechanics, and around the MaxEnt principle, found multiple successful applications in physics and elsewhere [11]. Many applications of MaxEnt are implemented in a classical setting. But there are also important applications to quantum problems. Starting with the works of Jaynes himself, the MaxEnt principle has been applied to quantum statistical mechanics in situations both at equilibrium and out of equilibrium. The MaxEnt principle has been applied even to the description of pure quantum states [12]. Generalizations of the MaxEnt principle, based on new information-entropic functionals [13], have also been explored and applied to a variety of problems, particularly in the field of complex systems [14]. Besides these information measures, there are other information-related quantities of physical relevance, such as Fisher's information measure [3]. Fisher's information was actually advanced before Shannon's [15], and represents a completely different concept [16]. It was introduced in the context of biology, but today constitutes an important tool in the study of diverse problems in physics (specially quantum physics) and other fields [3].

Another turning point in the story of scientists' gradual appreciation of the connection between physics and information was the formulation of Landauer's principle [17]. The principle says that there is a lower bound on the amount of energy that has to be dissipated each time that a bit of information is erased in a computing device. The minimum amount of energy that has to be dissipated is equal to $kT \ln 2$, where k is Boltzmann's constant and T is the absolute temperature at which the computer device works. Landauer's discovery established a direct and concrete connection between the concept of information and physical quantities such as energy and temperature. Landauer's principle constitutes a strong evidence that information has physical reality. This is nicely summarized in Landauer's famous motto "*information is physical*" [18]. Landauer principle has been the focus of intense research activity, and has been extended and generalized in diverse directions (see, for instance, [19] and references therein). Interest in the connections between physics and information increased substantially with the discovery that quantum mechanics allows for novel and highly counter-intuitive ways of transmitting and processing information. Some of the firsts steps in this direction were taken by Benioff [20], Feynman [21], and Deutsch [22], starting the field of quantum computation [6]. Around the same time, a striking feature of quantum information, encapsulated in Wootters and Zurek's quantum no-cloning theorem, was discovered [23]. These developments converged with other lines of inquiry going back to the works by Einstein, Podolsky and Rosen, and by Schroedinger, in the 1930s, on quantum nonlocality and entanglement, that together with later developments by von Neumann and by Bohm on hidden variables theories, led eventually to the discovery of Bell's inequalities (nice discussions on these developments can be found in [24]), and to the identification of quantum entanglement as one of the (if not *the*) most fundamental features distinguishing the quantum mechanical description of Nature from the classical one [25]. Afterwards, in the XXI century, the field of quantum information flourished and grew into myriad different directions [2], which are impossible to describe in this short note. Let us just mention that subjects central to the field of quantum information, such as quantum entanglement, are regarded by some researchers as key ingredients for understanding basic aspects of physics, such as the origin of gravity, Einstein's field equations, and the very structure of space–time [26].

2. Physics of information, quantum mechanics, and the future

The above are only a few highlights (corresponding particularly to the early steps) of the exploration of the connection between physics and information-related concepts. Even if summarized in a sketchy and incomplete fashion, they serve to illustrate a basic feature of these lines of inquiry. Research into the *physics of information*, which is here defined in a broad sense, as comprising all parts of physics, classical and quantum, where information-theoretical concepts play a central role, has two complementary facets. On the one hand, it investigates the capabilities allowed, and the limitations imposed, by the fundamental laws of Nature for the construction and operation of devices that transmit or process information [27–29]. A deep understanding of these issues may lead to revolutionary advances in information technologies, such as those expected from the fields of quantum communication and quantum computation. On the other hand, ideas and methods inspired in information theory help to achieve a deeper understanding of physics itself, as illustrated by Jaynes application of information theory to statistical mechanics [9, 10]. Rather than being in opposition, the two facets of the physics of information complement and stimulate each other. The friendly coexistence of the two facets reminds us of a famous quote by Poincare: "*I do not say: science is useful, because it teaches us to construct machines. I say: machines are useful because in working for us, they will some day leave us more time to make science. But finally it is worth remarking that between the two points of view there is no antagonism, and that man having pursued a disinterested aim, all else has been added unto him*" [30].

Research into the physics of information, including in particular the physics of quantum information, permitted the discovery of unexpected connections between apparently unrelated areas of science. New connections were established between different areas within physics, and also between physics and other sciences. As an illustration of the first kind of connections, we can mention that ideas related to Fisher's information suggested new connections between Schroedinger wave equation and Boltzmann transport equation [31]. With regards to the relationship between physics and other sciences, the physics of information is nowadays establishing profound relations between physics and biology [32]. The physics of information provides a set of theoretical and mathematical tools that constitutes a conceptual bridge between physics and biology. These developments, inextricably linked to the field of complex systems, include new theoretical ideas that affect all branches of biology. The study of consciousness constitutes perhaps the most remarkable example [33]. Until recently, the theory of consciousness was regarded as a subject that was outside the reach of scientific inquiry or, at least, outside the reach of a scientific treatment based on mathematically well-defined concepts, and amenable of quantitative experimental research. Although scientists, including physicists [34], have been interested in the phenomenon of consciousness for a long time, with psychologists and neuroscientists making a wealth of fascinating qualitative empirical discoveries, theoretical research into consciousness was largely regarded as a field of study for philosophers. The situation has changed dramatically in the last few years. Using ideas closely related to the physics of information, scientists are for the fists time attempting a mathematically-based theory of consciousness that might generate quantitative experimental predictions (see [33] and references therein). Most advances in the application to biology of methods or ideas related to the physics of information have been developed in a classical setting, but quantum mechanical aspects are starting to be explored in the new field of quantum biology [35]. There are even some intriguing hints suggesting that there might be connections between the phenomenon of consciousness and some basic aspects of quantum physics, such as the special and privileged role played in physics by the position observable [33].

The central role that the concept of information is gaining in physics raises some intriguing questions that deserve close scrutiny. The concept of information is, in a sense, a human-centered concept. After all, information theory was created to address engineering problems related to communication technology. We humans are the ones who care about information. Why should Nature care about information? Does nature care about information? These are perhaps naive questions. But we find it perplexing that a concept developed to address purely human needs turns out to be essential to understand the fabric of Nature at its deepest level. In this regard, one may also find intriguing that information-theoretical in physics reached their prominent role in physics precisely at a stage of human history, the “digital age”, when information technology became the most prominent technological feature of human life. This is probably not a coincidence. The question is, do we nowadays tend to interpret the laws of Nature in information-theoretical terms, and adopt the computer as our technological metaphor for natural systems and processes [29], because we are all the time using computers (particularly iPhones, around which the life of many revolve)? In other periods of History, the most advanced or sophisticated technological devices were also adopted as metaphors for Nature. In early modern times the metaphor was the clock. Today it is the computer. Is it going to be replaced by another metaphor in the future? We cannot know. From history, however, we learn that some of the insights gained from the old clock metaphor are still valuable. They have been incorporated, in terms compatible with the computer metaphor, to the law of conservation of information [5, 36–38]. This law says that at the most basic level the time evolution of physical systems preserves information. The conservation of information is one of the most fundamental laws of physics [5]. It is more profound and rich than the concepts embodied in the clock-metaphor. It holds both at the classical and at the quantum-mechanical levels, and its implications are manyfold. For instance, the quantum no-cloning theorem is a consequence of the law of conservation of information. Some basic aspects of quantum mechanical measurements, that until recently were presented in textbooks as part of the postulates of quantum mechanics, can actually be derived from the conservation of information [37].

Coming back to the question of which will be the future technological metaphor for Nature, it may happen that no metaphor will ever replace the computer one. It is conceivable that the computer is the ultimate metaphor for Nature because, in a sense, it is a universal metaphor. A universal Turing machine can compute or simulate anything that can be computed or simulated by a mechanical device. Consequently, as technological metaphors go, there may be nothing beyond the computer. And, concomitantly, the deepest description of Nature may admit its most adequate formulation in terms of ideas and concepts from computer science and information theory. Time will tell.

The dominant role that the physics of information, and specially quantum information, plays today manifest itself in various ways. For instance, in the number and geographical spread of researchers working in quantum information. Towards the end of the XX century, most research on quantum information was concentrated in a few countries. In many countries, there was still no activity in the field, or the field was just starting. Even in some countries with large and highly-developed research communities, and with big economies, the researchers working in quantum information were still very few. Today the situation is completely different. In all corners of the world, there are research groups enthusiastically exploring the many facets of quantum information, and making valuable contributions. Quantum information is nowadays a well established research field. The heroic days of the pioneers are over. This does not mean that the days of discovery are over. On the contrary, each new development generates new questions: research opportunities seem to be better than ever. And it may be the case that the best is yet to come.

Acknowledgements

We acknowledge partial financial support from ANID, grant MEC80190056, Chile. Additionally, one of us (S.C.) would like to thank Núcleo de Investigación No. 2–Sistemas Complejos en Ciencia e Ingeniería–UCN-VRIDT 042/2020, for the scientific support.

Conflict of interest

The authors declare no conflict of interest.

Author details

Sergio Curilef[1*†] and Angel Ricardo Plastino[2†]

1 Departamento de Física, Universidad Católica del Norte, Antofagasta, Chile

2 CeBio y Departamento de Ciencias Básicas, Universidad Nacional del Noroeste de la Provincia de Buenos Aires, UNNOBA, Conicet, Junin, Argentina

*Address all correspondence to: scurilef@ucn.cl

† These authors contributed equally.

References

[1] P. Davies and N. H. Gregersen. Information and the Nature of Reality. Cambridge University Press; 2014.

[2] O. Lombardi, S. Fortin, F. Holik, and C. Lopez, editors. What Is Quantum Information? Cambridge University Press; 2017.

[3] B. R. Frieden. Science from Fisher Information: A Unification. Cambridge University Press; 2004.

[4] H. S. Leff and A. F. Rex, editors. Maxwellâ's Demons 2: Entropy, Classical and Quantum Information, Computing. Bristol, Philadelphia: Institute of Physics Publishing; 2003.

[5] L. Susskind and J. Lindesay. An Introduction to Black Holes, Information, and the String Theory Revolution: The Holographic Universe. World Scientific; 2004.

[6] M. A. Nielsen and I. L. Chuang. Quantum Computation and Quantum Information. Cambridge: Cambridge University Press; 2000.

[7] C. E. Shannon. A mathematical theory of communication. Bell Syst. Tech. J. 1948;27:379–423.

[8] C. E. Shannon and W. Weaver. The Mathematical Theory of Communication. Urbana, IL: University of Illinois Press; 1949.

[9] E. T. Jaynes. Information theory and statistical mechanics. Phys. Rev. 1957; 106:620.

[10] E. T. Jaynes. Information theory and statistical mechanics. II. Phys. Rev. 1957; 108:171.

[11] S. Pressé, K. Ghosh, J. Lee, and K. A. Dill. Principles of maximum entropy and maximum caliber in statistical physics. Rev. Mod. Phys. 2013;85:1115-1141.

[12] A. R. Plastino and A. Plastino. Maximum entropy and approximate descriptions of pure states. Phys. Lett. A 1993;181:446-449.

[13] C. Beck. Generalised information and entropy measures in physics. Contemp. Phys. 2009;50:495.

[14] C. Tsallis. Introduction to Nonextensive Statistical Mechanics – Approaching a Complex World. New York: Springer; 2009.

[15] R. A. Fisher. Theory of statistical estimation. Math. Proc. Cambridge Philos. Soc. 1925;22:700.

[16] A. R. Plastino and A. Plastino. What's the big idea? Cramer-Rao inequality and Rao distance. Significance 2020;17:39.

[17] R. Landauer. Irreversibility and heat generation in the computing process. IBM J. Res. Dev. 1961;5:183.

[18] R. Landauer. Information is physical. Phys. Today 1991;44:23.

[19] C. Zander, A. R. Plastino, A. Plastino, M. Casas, and S. Curilef. Landauer's principle and divergenceless dynamical systems. Entropy 2009;11: 586-597.

[20] P. Benioff. The computer as a physical system: A microscopic quantum mechanical Hamiltonian model of computers as represented by Turing machines. J. Stat. Phys. 1980;22: 563–591.

[21] R. P. Feynman. Simulating physics with computers. Int. J. Theor. Phys. 1982;21:467.

[22] D. Deutsch. Quantum theory, the Church-Turing principle and the universal quantum computer.

Proceedings of the Royal Society of London A 1985;400:97–117.

[23] W. K. Wootters and W. H. Zurek. A single quantum cannot be cloned. Nature 1982;299:802.

[24] J. S. Bell. Speakable and Unspeakable in Quantum Mechanics. 2nd ed. Cambridge University Press; 2004.

[25] I. Bengtsson and K. Zyczkowski. Geometry of Quantum States: An Introduction to Quantum Entanglement. Cambridge University Press; 2006.

[26] C. J. Cao and S. M. Carroll. Bulk entanglement gravity without a boundary: Towards finding Einstein's equation in Hilbert space. Phys. Rev. D 2018;97:086003.

[27] N. Margolus and L. B. Levitin. The maximum speed of dynamical evolution. Physica D 1998;120:188.

[28] S. Lloyd. Ultimate physical limits to computation. Nature 2000; 406:1047.

[29] S. Lloyd. Computational capacity of the universe. Phys. Rev. Lett. 2002;88: 237901.

[30] H. Poincare. The Value of Science. Dover; 1958.

[31] S. P. Flego, B. R. Frieden, A. Plastino, A. R. Plastino, and B. H. Soffer. Nonequilibrium thermodynamics and Fisher information: Sound wave propagation in a dilute gas. Phys. Rev. E 2003;68:016105.

[32] P. Davies. Does new physics lurk inside living matter? Physics Today 2020;73:34-40.

[33] M. Tegmark. Consciousness as a state of matter. Chaos, Solitons & Fractals 2015;76:238-270.

[34] E. Schroedinger. What Is Life? With Mind and Matter and Autobiographycal Sketches. Cambridge Universirty Press; 1996.

[35] Y. Kim, et al. Quantum biology: An update and perspective. Quantum Reports 2021;3:80-126.

[36] A. R. Plastino and A. Daffertshofer. Liouville dynamics and the conservation of classical information. Phys. Rev. Lett. 2004;93:138701.

[37] W. H. Zurek. Quantum origin of quantum jumps: Breaking of unitary symmetry induced by information transfer in the transition from quantum to classical. Phys. Rev. A 2007;76: 052110.

[38] C. Zander and A. R. Plastino. Fidelity measure and conservation of information in general probabilistic theories. EPL 2009;86:18004.

Section 2

Fundamentals

Chapter 2

Ontology in Quantum Mechanics

Gerard 't Hooft

Abstract

It is suspected that the quantum evolution equations describing the micro-world as we know it are of a special kind that allows transformations to a special set of basis states in Hilbert space, such that, in this basis, the evolution is given by elements of the permutation group. This would restore an ontological interpretation. It is shown how, at low energies per particle degree of freedom, almost any quantum system allows for such a transformation. This contradicts Bell's theorem, and we emphasise why some of the assumptions made by Bell to prove his theorem cannot hold for the models studied here. We speculate how an approach of this kind may become helpful in isolating the most likely version of the Standard Model, combined with General Relativity. A link is suggested with black hole physics.

Keywords: foundations quantum mechanics, fast variables, cellular automaton, classical/quantum evolution laws, Stern-Gerlach experiment, Bell's theorem, free will, Standard Model, anti-vacuum state

1. Introduction

Since its inception, during the first three decades of the 20th century, quantum mechanics was subject of intense discussions concerning its interpretation. Since experiments were plentiful, and accurate calculations could be performed to compare the experimental results with the theoretical calculations, scientists quickly agreed on how detailed quantum mechanical models could be arrived at, and how the calculations had to be done.

The question what the intermediate results of a calculation actually tell us about the physical processes that are going on, remained much more mysterious. Opinions diverged considerably, up to today, one hundred years later.

The historical events that led to this situation are well-known, and have been recounted in excellent reports [1]; there is no need to repeat these here extensively. It was realised that all oscillatory motion apparently comes in energy packets, which seem to behave as particles, and that the converse should also be true: all particles with definite energies must be associated to waves. The original descriptions were somewhat vague, but the year 1926 provided a new landmark: Erwin Schrödinger's equation [2]. Originally, the equation was intended to describe just one particle at the time, but soon enough it was found how to extend it to encompass many particles that may be interacting.

Indeed, in his original paper, Schrödinger went quite far in discussing Hamilton's principle, boundary conditions, the hydrogen atom and the electromagnetic transitions from one energy level to an other. One extremely useful observation was made by Max Born [3]: the absolute square of a wave function, at some spot in position space, must simply stand for the *probability* to find the particle there. This made a lot of sense, and it was rightly adopted as a useful recipe for dealing with the equation.

But then, many more questions were asked, many of them very well posed, but the answers sound too ridiculous to be true, and, as I shall try to elucidate, they *are* too ridiculous to be true. I am not the only scientist who feels taken aback by the imaginative ideas that were launched, ranging from the role of 'guiding pilot' adopted by the wave function [4] to steer particles in the right direction, to the idea that infinitely many 'universes' exist, all forming parts of a more grandiose concept of 'truth' called 'multiverse' or 'omniverse', an idea now known as the 'many worlds interpretation' [5, 6].

In contrast, an apparently quite reasonable conclusion was already reached in discussions among scientists in the 1920s, centred around Niels Bohr in Copenhagen, called the 'Copenhagen Interpretation'. They spelled out the rules for formulating what the equations were, and how to elaborate them to make firm predictions. Indeed, we know very well how to use the equation. The properties of atoms, molecules, elementary particles and the forces between all of these can be derived with perplexing accuracy using it. The way the equation is used is nothing to complain about, but what exactly does it say?

Paul Dirac for instance, advised not to ask questions that cannot be answered by any experiment; such questions cannot be important. We know precisely how to use Schrödinger's equation; all that scientists have to do is insert the masses and coupling parameters of all known particles into the equation, and calculate. What else can you ask for? Many of my colleagues decided to be strictly 'agnostic' about the interpretation, which is as comfortable a position to take as what is was for 19th century scientists to stay 'agnostic' about the existence of atoms.

The Copenhagen verdict was:

> *"There are many questions whose answers will not be in the range of any experiment to check; there will be no unanimous agreement on the interpretation of the equations, so stop asking."*

The present author accepts all conclusions the Copenhagen group had reached, except this last one. It will be important to ask for models that can elucidate the events that take place in an experiment. We do wish to know which sensible models can in principle explain the Schrödinger equation and which will not.

What happens to its wave function when you actually observe a particle? What does it mean if the Schrödinger equation suggests that interference takes place between different possible paths a particle can take? Those questions I can now answer, but others are still way out of reach: the masses and coupling parameters of the elementary particles have been determined by experiment, but we do not have acceptable theories at all to explain or predict their values. If the history of science is something to be taken to mind, it may be that asking unconventional questions will lead to better insights.

The Schrödinger equation is simple and it works, but some of the explanations why it works seem to get the proportions of a Hieronymus Bosch painting. This does not sound right. Almost a full century has passed since the equation was written down, and we still do not know what or whom to believe, while other scientists get irritated by all this display of impotence [7]. Why is it that we still do not agree?

I think I know some of the answers, but almost *everyone* disagrees with me. I have reached the conclusion that quantum mechanics indeed describes a completely deterministic world. Admittedly, I will leave some questions unanswered. The origin of the symmetries exhibited by the equations is not well understood. More advanced mathematics will have to be employed to answer such questions, as will be explained. Sharpening the scope of my claim, the point is that there is no mystery with quantum mechanics itself. Just leave questions concerning symmetries aside for the time being. In contrast with what others proclaimed, there is no logical conflict. This will be explained (Section 5).

What *are* those masses and coupling strengths? Do particles exist that we have not yet been able to detect? Isn't it the scientist's job to make predictions about things we have not yet been able to unravel? These are questions that are haunting us physicists. We have arrived at a splendid theory that accounts for almost anything that could be observed experimentally. It is called the *Standard Model* of the subatomic particles. But this model also tells us that particles and forces may exist that we could not have detected today. Can we produce any theory that suggests what one might be able to find, in some distant future? And as of all those particles and forces that we do know about, is there a theory that *explains* all their details?

Today's theories give us little to proceed further from where we are now. The Standard Model explains a lot, but not everything. This is why it is so important to extend our abilities to do experiments as far as we can. Recently, audacious plans have been unfolded by the European particle physics laboratory CERN, for building a successor of its highly successful Large Hadron Collider (LHC). While the experimental groups working with the LHC have provided for strong evidence supporting the validity of the Standard Model up to the TeV domain, theoreticians find it more and more difficult to understand why this model can be all there is to describe what happens further beyond that scale. There must be more; our present theoretical reasoning leads to questions doubting the extent to which this model can be regarded as 'natural' if more of the same particles at higher energies are allowed to exist, while the existence of totally new particles would be denied.

Inspired by what historians of science are telling us about similar situations in the past history of our field, investigators are hoping for a 'paradigm shift'. However, while it is easy to postulate that we 'are doing something wrong', most suggestions for improvement are futile; suggesting that the Standard Model would be 'wrong' is clearly not going to help us. The 'Future Circular Collider' is a much better idea; it will be an accelerator with circumference of about 100 km, being able to reach a c.m. collision energy of 100 TeV. The importance of such a device is that it will provide a crucial background forcing theoreticians to keep their feet on the ground. if you have a theory, it better agree with the newest experimental observations.

2. The generic realistic model

The central core of our theory consists of a set of models whose logic is entirely classical and deterministic. Deterministic does not mean *pre*-deterministic: there is no shortcut that would enable one to foresee any special feature of the future without performing extremely complex simulation calculations using the given evolution laws. There is no 'conspiracy'. Also, we do not take our refuge into any form of statistics. The equations determine exactly what is happening. Of course we do not know today exactly what the equations are, but we do assume them to exist.

The equations will be *more precise* even than Newton's equations for the motion of the planets. Newton's equations are given in terms of variables whose values are determined by real numbers. But, in practice, it is impossible to specify these numbers with infinite precision, and consequently, *chaos* takes place: it is fundamentally impossible, for instance, to predict the location of the dwarf planet Pluto, one billion years from now, because such a calculation would require the knowledge of the locations and masses of all planets in more than 20 digits accuracy today [8]. That's a tiny fraction of a micron for Pluto's orbit. Following Pluto during the age of the universe would require accuracies beyond 2000 digits, much tinier margins than the Planck length. To describe Pluto's position exactly would require an infinite number of decimal places to be rigorously defined.

Our deterministic theory will be formulated in terms of integer numbers only, which can be defined exactly without the need of infinitely many decimal places. This kind of precision in defining theories may well be what is needed to understand quantum mechanics.

For simplicity, we imagine a universe with finite size and finite time. As for their mathematical structure, all deterministic models are then very much alike. All finite-size discrete models must have finite Poincaré recursion times. There will be different closed cycles with different periods, see **Figure 1**. Counting these cycles, one finds that the rank of a cycle is physically a conserved quantity, almost by definition. For simplicity, we constrain ourselves to *time reversible* evolution laws, although it is suspected that one might be allowed to relax this rule, but then the mathematics becomes more complex.

We now emphasise that the evolution law of such a deterministic system can be exactly described in terms of a legitimate, conventional Schrödinger equation. We say that quantum mechanics is a *vector representation* of our model: every possible state the system can be in is regarded as a vector in the basis of Hilbert space. This set of vectors is orthonormal. The classical evolution law will send any of these vectors into an other one. Since these vectors are all orthonormal and since the evolution is time-reversible, one can easily prove that the evolution matrix is unitary. It contains only the numbers 1 and 0. There is only one 1 in each row and in each column; all the other entries are 0, from which unitarity follows.

By diagonalising this matrix, one finds all its eigenvectors and eigenvalues. Within one cycle, the eigenvalues of $U(t)$ are $e^{-2\pi int/T}$, where t is time, T is the period, and n is an integer. The formal expression for the eigenvectors is easily obtained:

$$ {}^{\text{ont}}\langle k|n\rangle^{E} = \frac{1}{\sqrt{N}} e^{-2\pi ink/N} \ , \tag{1} $$

where $|k\rangle^{\text{ont}}$ are the ontological states, labelled by the integer k, and $|n\rangle^{E}$ are the energy eigenvectors. We read off in the basis formed by the states $|n\rangle^{E}$ that the Hamiltonian takes the values.

$$ H_{nm} = 2\pi n\,\delta_{nm}/T. \tag{2} $$

At first sight, this does not look like quantum mechanics; the series of eigenvalues (2) seems to be too regular. In [9] it was proposed to add arbitrary additive

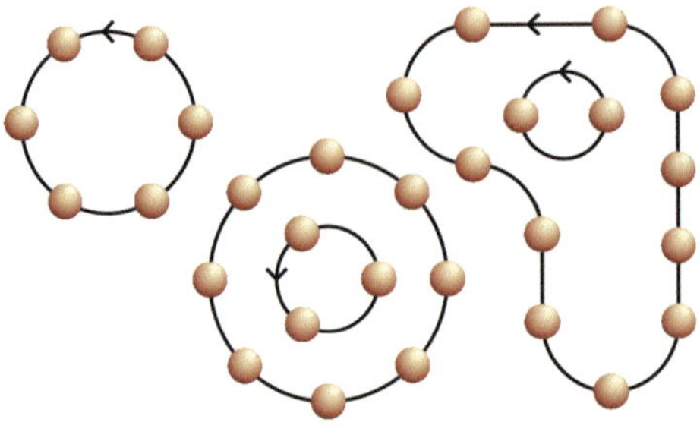

Figure 1.
Generic evolution law for a realistic model with different periodicities. In this example we see 5 cycles, with ranks 2, 3, 6, 8 and 11.

energy renormalization terms, depending on the cycle we are in, but the problem is then still that it is difficult to see how this can reproduce Hamiltonians that we are more familiar with. The energy eigenstates seem to consist of large sequences of spectral lines with uniform separations. A more powerful idea has been proposed recently [10, 11]. More use must be made of locality. We wish the Hamiltonian to be the sum of locally defined energy density operators,

$$H = \sum_{\vec{x}} \mathcal{H}(\vec{x}) \ . \tag{3}$$

Now this is really possible. The price to be paid is to add *fast fluctuating, localised variables*, called 'fast variables' for short. They replace the vague 'hidden variables' that were introduced in many earlier proposals [12].

The fast variables, $0 \le \varphi_i(\vec{x}) < 2\pi$, are basically fields that rapidly repeat their values with periodicities $T_i(\vec{x})$, which we choose all to be large and mostly different. To reproduce *realistic* quantum mechanical models, we need these periods to be considerably shorter in time than the inverse of the highest energy collision processes that are relevant.

To a good approximation, the fast variables will be non-interacting. This means that the energy levels will take the form $E = 2\pi\sum_{i,\vec{x}} n_i(\vec{x})/T_i$, where the n_i are all integer, and it implies that there is one ground level, $E_0 = 0$, while all excited states have energies $E \ge 2\pi/T_i$. Clearly, our conditions on the fast variables were chosen such that their excited energy levels exceed all energy values that can be reached in our experiments.[1]

Note that energy is exactly conserved. Therefor we may assume that, if an initial state is dominated by the state $|E = 0\rangle$, it will stay in that state.

Now consider the quantum model that we wish to mimic. Let that have a basis of N states, $|\alpha\rangle, |\beta\rangle, \cdots$, with $1 \le \alpha, \beta, \cdots < N$, to be called the *slow* variables. Their interactions are introduced as *classical* interactions with the fast variables, as follows:

Two states $|\alpha\rangle$ and $\beta\rangle$ are interchanged whenever the fast variables in the immediate vicinity of states α and β simultaneously cross a certain pre-defined point on their (fast) orbits.

Here, the 'vicinity' must be a well-defined notion for these states. In the case of non-relativistic particles, it means that we defined the states as the particle(s) in the coordinate representation $|\vec{x}(t)\rangle$. In the relativistic case we take the basis of states specified by the fields $\phi_i(\vec{x})$. This does imply that, in both cases, we regard the particles and/or fields to undergo exchange transitions that eventually will generate the desired Schrödinger equation or field equations.

One can describe these classical interchange transitions in terms of a 'quantum' perturbation Hamiltonian.

$$H^{\rm int} = \frac{\pi}{2} \sum_{\alpha,\beta,s} \sigma_y^{[\alpha,\beta]} \, \delta_{\varphi_\alpha, \varphi_\alpha^{(s)}} \, \delta_{\varphi_\beta, \varphi_\beta^{(s)}} \ , \tag{4}$$

where $\sigma_y^{[\alpha,\beta]}$ is one of the three Pauli matrices $\sigma_x, \sigma_y, \sigma_z$, acting on the two-dimensional subspace spanned by the two states $|\alpha\rangle$ and $|\beta\rangle$.

---

[1] More precisely, we talk of energies that can be associated to single quantum particles at isolated points in space–time.

Some special points on the orbits of the fast variables φ_α and φ_β will be indicated as $\varphi_\alpha^{(s)}$ and $\varphi_\beta^{(s)}$. If the fast variables φ_α and φ_β reach their special positions *simultaneously* then the corresponding classical states $|\alpha\rangle$ and $|\beta\rangle$ are interchanged.

In Eq. (4), we used a discretised notation, where the time unit is chosen such that it is the time needed to advance the fast variables by only one step in their (discretised) orbits. One may check that the factor $\pi/2$ is crucial to guarantee that, if the special point is reached, the equation.

$$e^{-\frac{\pi i}{2}\sigma_y} = -i\sigma_y = \begin{pmatrix} 0 & -1 \\ 1 & 0 \end{pmatrix} \tag{5}$$

describes a classical interchange, without generating superpositions. The minus sign is unavoidable but causes no harm. We chose the Pauli matrix σ_y because, when combined with the factor i in the Schrödinger equation, the wave function will be propagated as a *real*-valued quantity. One might desire to generate one of the other Pauli matrices also using classical physics. This can be done by adding a dummy binary variable, as described in [11] (the binary variable also propagates classically).

The Hamiltonian describing the evolution of the slow variables is now derived by assuming that the fast variables never get enough energy to go to any of their excited energy states. Their lowest energy states are $|0\rangle^E$ obeying

$$^{\text{ont}}\langle k|0\rangle^E = \frac{1}{\sqrt{N}} , \tag{6}$$

so that the expectation value of a Kronecker delta is

$$\left\langle 0|\delta_{\varphi_\alpha,\varphi_\alpha^{(s)}}|0\right\rangle = \frac{1}{N} , \tag{7}$$

where N is the number of points on the fast orbit of this variable.

Eq. (5) could als be used if we had only one Kronecker delta in Eq. (4), but this would cause exactly one transition during one period of the fast variable, which makes the effective Hamiltonian too large to be useful. Choosing two Kronecker deltas causes one transition only to take place after much more time, making the insertion (4) of the desired order of magnitude to serve as a contribution in the effective Hamiltonian of the slow variables.

By adding a large number of similar transition events in the orbits of all fast variables, causing transitions for all pairs of (neighbouring) slow variables, we can now generate any desired contributions to the effective Hamiltonian elements $H_{\alpha\beta}$ causing transition among the slow variables. The result will be.

$$H^{\text{int}}_{\alpha\beta} = \frac{\pi}{2}\sigma_y^{[\alpha,\beta]} \frac{N^{[s]}}{N_{[\alpha]}N_{[\beta]}} , \tag{8}$$

where the numbers $N_{[\alpha]}$ and $N_{[\beta]}$ are the total numbers of points on the orbits of the fast variables α and β, and the numbers N^s indicate the numbers of the special transition points on the donut formed by the orbits of the pair α, β.

We encounter the restriction that the matrix elements will come with rational coefficients in front. The fundamental reason for the coefficients to be rational is that, eventually, all discretised classical models have finite Poincaré recursion times. In practice one may expect that this problem goes away when, for realistic classical systems, the Poincaré recursion times will rapidly go to infinity.

We have now achieved the following. Let there be given any Hamiltonian with matrix elements $H_{\alpha\beta}$ in a finite-dimensional vector space, and given a suitably chosen basis in this vector space, preferably one where every state can be endowed with coordinates $\vec{x}$. Then we have defined slow variables $|\alpha\rangle$ describing the physical states, and we added fast variables whose excited states are beyond the reach of our experiments. We found classical interactions, prescribed as exchanges between the classical states, such that the effective Hamiltonian will approach the given one.

The system obeys the Schrödinger equation dictated by this Hamiltonian, and, by construction, all probabilities evolve as is mandated by the Copenhagen doctrine. The reader may ask how to obtain the diagonal elements of the Hamiltonian, and how to make it contain complex numbers. The answer is that these can also be generated by using an additional binary degree of freedom as mentioned above.

In principle, one could have used any orthonormal basis of states to be used in our construction, but in practice we would like to recover locality in some way. The demand of locality in the classical system implies that we should demand locality for the fast variables and the slow ones. This appears to be straightforward. For non-relativistic particles, one may use the basis of states defined by the position operators $\vec{x}$. In the relativistic case, one needs the field operators $\varphi_i\left(\vec{x}\right)$ and their quantum eigen states to start off with.

The theory we arrive at appears to be closely related to Nelson's 'stochastic quantum mechanics' [13]. We think our construction has a more solid mathematical foundation, explaining how the quantum entanglement arises naturally from the energy conservation law, associated to time translation invariance.

3. Symmetries and superpositions

The interpretation of the Schrödinger equation that we obtained is that it merely describes the evolution of the probability distribution for the slow variables, after averaging over the positions of the fast variables. The fast variables dictate the evolution, but they act too fast for us to observe this directly. The new thing in our procedure is that we have the choice to also describe the fast variables using quantum mechanics as a tool: fast and slow variables together go into a vector representation of what happens.[2]

In statistical treatments of moving variables, with well-determined evolution laws, it should be completely clear that the probability distributions of the final state are the result of our choices for the probability distributions of the initial states.

At first sight, the group of rotations can also be regarded as pure permutations, and, although the lattice structure of our local coordinate space $\vec{x}$ will be severely affected, one might suspect that our present understanding of physics comes from smearing the lattice back into a continuum; this may be a reasonable approach towards understanding rotational symmetry.

However, more severe problems arise if we consider the notion of spin in a particle. We need to take spin into account when analysing Bell's theorem. In the treatment displayed in the previous section, the spin variable of a particle would be a discretised variable s, with integer spacings ranging from $-S$ to S, where S is the

---

[2] Do keep in mind that the distinction between fast variables and slow variables is a feature of our simplified models, but possibly unnecessary in the real world. All variables are real, evolving classically according to the same or similar classical laws.

total spin quantum number, being integer or half-odd-integer. These would be promoted to the status of classical variables, and then we can set up exactly the right Schrödinger equation for particles like the Dirac particle. What happens with its ontological interpretation if we rotate that?

It is clear that, in this case, rotation transformations transform the 'real states' $|s\rangle$ into superpositions. In doing so, the rotation group can serve as the prototype of many symmetry considerations in quantum mechanics. How do we analyse the Stern-Gerlach experiment?

The superimposed states obey exactly the same Schrödinger equations as the basis elements do, and we had chosen the latter at will. So one possible answer could be: it does not matter which of the states we call real; there is no experiment to help us make the distinction. But this is debatable. The Stern-Gerlach experiment in its vertical orientation distinguishes particles with spin up from particles with spin down, these have different orbits. Remember that, in this chapter, we focus on going beyond the usual statistical interpretation of quantum mechanics, aiming at a description of pure, real states. The only accepted probabilities are 1 and 0.

The real physical states we work with form a basis of Hilbert space, and the equations we work with ensure that any state that starts off being real, occurring either with probability 1 or with probability 0, continues to be a real state forever. This must also hold for any Stern-Gerlach set-up or any of the other paradoxical contraptions that have been proposed over the years. Real state in = real state out. This was called the 'law of ontology conservation' [14].

At first sight it seems that a Stern-Gerlach experiment, after a rotation over an arbitrary angle, turns into a superposition of several real states. This is true in the mathematical sense. It is the easiest way to visualise what the rotation group stands for. However, if we physically rotate a Stern-Gerlach experiment, by undoing and re-arranging nuts and bolts, We do something else. The new experiment again goes into one of the realistic states; the nuts and bolts also go into new physical states, so this is not quite the same kind of rotation.

Notice however, that if a particle with spin leaves one Stern-Gerlach instrument and continues its way in an other, rotated, device, then, as we know from standard quantum mechanics, it emerges in a superposition, or more accurately, in a probabilistic distribution. Where does this stochastic behavior come from? What happens if we do interference experiments with the various emerging beams of particles?

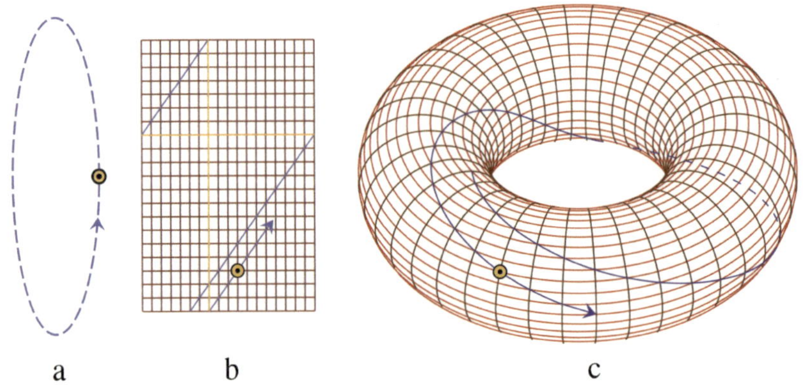

Figure 2.
The periodic orbits of the fast variables. Points where interactions take place are indicated. If these occur in the orbit of a single fast variable (a), they will be difficult to miss, but in the case of two or more (b), the special points will be hit much less frequently, so that the interactions become slow. The orbit takes the shape of a (multidimensional) torus (c).

Apparently, there are other variables that play a role. We can blame the fast variables for this. The fast variables for the rotated device do not coincide with the previous fast variables. In specifying the state of the particle in the first device, we forgot to observe where exactly the fast variable was. We couldn't observe this, as it was moving too fast. The transformation (in this case that is the rotation), formally involved the excited energy modes of the fast variables. In practice, we know that the energies of the quantum particles in both devices are too low to detect the excited modes, but in formulating the interactions, using the special points in **Figure 2b** and **c**, the excited modes do play a role because the interaction points are localised.

From these considerations, we claim that whatever is left of the various paradoxes should be nothing to worry about.

4. On Bell's theorem

Yet, this conclusion is often criticised. To set the stage, let us recapitulate J.S. Bell's theorem [15]: a source is constructed that emits two entangled photons simultaneously. Such sources exist, so no further justification of its properties is asked for. If $\pm z$ is the direction of the photons emitted, then the helicities are in the xy direction. Entanglement here means that the 2-photon state is.[3]

$$|\psi\rangle_{\text{source}} = \frac{1}{\sqrt{2}}(|0\,0\rangle + |1\,1\rangle) \ , \tag{9}$$

where the 0 stands for the x polarisation and 1 stands for the y polarisation. Alice and Bob use polarisers to analyse the photons they see. Alice rotates her polariser to an angle a and Bob chooses an angle b, and these choices are assumed to be totally independent. The two photons "do not know" what the angles a and b are; they are assumed to emerge with a polarisation angle λ. According to the usual view of what hidden variables are, the probability that both Alice and Bob are detecting their photon is written as $P(a,b)$; the probability that a photon with orientation λ is detected by Alice is assumed to be $p_A(\lambda,a)$, and the probability that Bob makes an observation is written as $p_B(\lambda,b)$. One then writes.

$$P(a,b) \stackrel{?}{=} \int \mathrm{d}\lambda \cdot \rho(\lambda) \cdot p_A(a,\lambda) \cdot p_B(b,\lambda). \tag{10}$$

All probabilities p and P are assumed to be between 0 and 1. $\rho(\lambda)$ is also positive and integrates to one. Bell would argue that this expression should apply to theories such as ours, simply by merging the fast variables $\varphi_i\left(\vec{x}\right)$ with the parameter λ.

Figure 3 shows what assumptions go in Eq. (10). It seems to be obvious that observers in regions 1 and 2 may choose any setting a and b to identify their photon.

Writing $\overline{a} = a \pm 90°$, photons obey:

$$P(\overline{a}) + P(a) = 1 \ . \tag{11}$$

The correlation between Alice's and Bob's measurement is then written as.

---

[3] Alternatively, a source emitting two spin 1/2 particles could be used. The angles of the polarisation will then be twice the angles of the photon orientation discussed here, and there will be other modifications due to the fact that these particles are fermions.

Figure 3.
Bell's definition of locality. 1 and 2 are two small regions of space–time, space-like separated. "Full specification of what happens in region 3 makes events in 2 irrelevant for predictions about 1 if local causality holds".

$$E(a,b) = P(a,b) + P(\bar{a},\bar{b}) - P(a,\bar{b}) - P(\bar{a},b) \ . \tag{12}$$

Standard quantum mechanics allows one to choose the entangled photon state (9) as if it is oriented towards either Alice or Bob, since it is rotation independent. The outcome is then.

$$E_{\text{quant}}(a,b) = 2\cos^2(a-b) - 1 = \ \cos 2(a-b) \ . \tag{13}$$

In the fashionable hidden variable language, Eq. (10) is assumed to be valid, which implies that the photon must take care of giving Alice and Bob their measurement outcome whatever their choices a and b are, and these outcomes are found to obey the CHSH inequality [16], derived directly from Eq. (10). One then finds that Eq. (5) conflicts with Eq. (10). There is a mismatch of at least a factor $\sqrt{2}$, realised when $|a-b| = 22.5^\circ$ or 67.5°.

Several 'loopholes' were proposed, having to do with the limited accuracy of the experiments, but these will not help us, since we claim that our theory exactly reproduces quantum mechanics, and therefore Eq. (13) should be reproduced by our theory. It violates CHSH. How can this happen?

Our short answer is that we have a classically evolving system that exactly reproduces the probability expressions predicted by the Schrödinger equation, including Eq. (13), in a given basis of Hilbert space. The model is local and allows for any initial state; it does not require any kind of 'conspiracy' or 'retrocausality', or even non-locality.

This should settle the matter, but it is true that the violation of the CHSH inequality is quite surprising.

The difficulty resides in assumption (10). Bell derives it directly from causality. If no signal can travel from the space–time point where Alice does her measurement to the point where Bob does his experiment, and *vice versa*, then Eq. (10) just follows. Nevertheless, an assumption was made.

It amounts to the statement that the variables $\lambda, a,$ and b, are mutually independent. However, in [9], we computed the minimal non-vanishing correlations between the angles a, b and λ that could reproduce the quantum expression (13) exactly. We found[4]:

$$P(a,b,\lambda) = C|\sin 2(a+b-2\lambda)| \ , \tag{14}$$

where C normalises the total probability to one (its value depends on the integration domain chosen). This expression shows a non-vanishing 3-variable correlation, without any 2-variable correlations as soon as one averages over any of the three variables. An equation such as (14) should replace (10).

---

[4] This outcome is model dependent, and if we choose the model to be physically more plausible, the correlations become even stronger.

One can read this to mean that the settings a and b have an effect on λ, but one can also say that the choice of λ made by the photon, affected the settings chosen by Alice and Bob.[5] Perhaps the best way to interpret this strange feature is that it be an aspect of information: the fact that the fast variables occupy all states in their orbits with equal probabilities is expressed by saying that they live in their energy ground states. The choice of the phases here is a man-made ambiguity that may propagate backwards in time. It is not an observable 'spooky signal', since nothing propagates backwards in time in the classical formulation.

When we say that the photons (together with the fast variables) 'affect' the settings chosen by Alice and Bob, it implies that Alice and Bob have no 'free will'. Of course they haven't, their actions are completely controlled by the equations. We can't change setting a without changing what happens in region 3 of **Figure 3**.

It is important then to realise that our theory is *not* a theory about statistical distributions. If we include the fast variables, everything that happens in region 3, occurs with probability 1, or, if it does not happen, it has probability 0. There is no in-between. Remember that we reproduce the Schrödinger equation *in a given basis of Hilbert space*. The probabilities of the Schrödinger equation emerge exactly, but only if we start with the right basis elements.

We can add to this an important observation when the classical degrees of freedom are considered: *even a minute change of the setting a will require an initial state in* **Figure 3**, *region 3 that is orthogonal to what it was before that adjustment.* This is because the settings are classically described. The required rotation of the fast variable erases the information as to where its transition point was located (see **Figure 2**).

We note that this aspect of our scenario implies the absence of 'free will' for Alice and Bob in choosing their settings. Alice and Bob are forced to obey classical laws, such that the rule *ontology in* = *ontology out* is obeyed. The same can be said of Schrödinger's cat. Eventually, what we see when inspecting the cat is its classical behaviour. Only after adding the (in practice invisible) fast variables, we can perform a basis transformation to quantum states to say that the cat is superimposed. The statement belongs in the world of logic generated by the vector representation, but means nothing as long as we hold on to the classical description.

5. Where are the fast variables and the slow variables in the standard model?

At first sight we may seem to be a long way from describing quantum field theories such as the Standard Model. In principle, one may expect something resembling a cellular automaton, where we may be able to project the various field variables as data on a cellular lattice. However, as described in Section 3, we have to deal with the question how continuous and discrete symmetry patterns, essential for the Standard Model to work, can be introduced. As is well-known, once we have all local and global symmetries in place, the entire Standard Model is almost fixed, with only a few dozen interaction parameters to be determined. We make a gentile attempt at finding some sign posts that could indicate to us where to start.

In a very important paper [17], F. Jegerlehner describes the Standard model as a minimalistic outcome of an algebraic structure whose basic interaction properties

[5] This then would be an example of the 'butterfly effect'. It is not as crazy as it sounds. As soon as we include the fast variables in the discussion, the dynamics becomes invariant under time reversal, and the statement that a later photon is correlated with settings chosen earlier is then not strange at all.

are essentially natural near its ultimate cut-off scale, the Planck length, except that the Higgs field self coupling happens to vanish, or almost vanish, at that scale. It seems as if the universe is metastable, or perhaps just at the edge of stability. When we scale towards the TeV scale, using the renormalization group equations, one discovers that the Higgs self-coupling slowly grows towards its present value, and this appears to explain the recently observed Higgs mass remarkably well.

There are important new questions that may be raised in connection with the present work. One is where in the cellular automaton this copious algebra is generated; and of course we want to know how any kind of fast oscillating variables can arise. Previously, this author was just thinking of very heavy virtual particles such as the vector bosons that represent the remaining grand unification symmetry, but there is a problem with that: as described in Section 2, **Figure 2**, the dynamical fast variable must have the geometry of a multi-dimensional torus, whereas fields have a more trivial topological structure if they indeed form vector representations of the unifying algebra, see Section 2.

This perhaps can be done better [18]. The general philosophy that might be useful here starts from a fundamental observation. Fields that describe data at the Planck scale, can only propagate as fields at much more conventional scales (from milli-electronVolts to nearly a TeV), if there is a mechanism that prevents them from obtaining effective mass terms. To be precise, the dynamical field equations must allow them to be shifted by a constant with only minor effects on the energy of the state. At our scale of physics (to be referred to as the SM scale), fields can be shifted in any way, depending on space and time, such that energies also change within the energy domain of our SM scale. This means that the effective mass term must be at the SM scale. When we move towards the Planck scale, this mass term must rapidly approach to zero. Physically, the only mechanism that can do this is the Goldstone mechanism:

Only if a field effectively describes a symmetry transformation, and if, at the Planck scale, our world is invariant under this symmetry transformation, then we can understand how this field can propagate all the way to the SM scale.

Since the Standard Model has a rich spectrum of possible fields (fermionic and bosonic), this would force us to suspect that each of these fields must represent a symmetry transformation under which the Planck-scale theory is either exactly invariant (when the mass term vanishes) or invariant in a very good approximation (when the mass is of SM scale or smaller). Indeed, this should also hold for the fermionic fields, and this points towards supersymmetry at the Planck scale.

In short, every field component in the SM represents a generator for an almost exact symmetry of the Planck scale model. If we would be dealing with only scalar fields in the Standard Model, this would give us all the symmetry transformations, including estimates on how well the system is invariant.

Unfortunately, the real situation will be a lot more complicated. We have fermionic fields that transform as spinors under rotation, and we have vector fields that themselves again obey local gauge symmetries. How do we deal with that? It would be a great assignment for a team of PhD students to design and elaborate a logically coherent mathematical scheme.

This scheme might eventually produce logical guidelines for setting up cellular automaton models in such a way that their behavior at SM scales indeed reproduce the SM. But this is not all. The resulting automaton will still be a *quantum* automaton. What we now need is a set of variables that can play the role of fast variables. These are fields, but they cannot live on a flat field-space, they must form toruses as in Section 2. Now it would be tempting to consider the gauge groups. All group parameters of the local gauge groups $SU(2), SU(3)$, and $U(1)$, form toroidal spaces or spheres. What's more, we know that the physical quantum states are invariant under these group transformations, so regardless their time-dependence, our world should be in the invariant state, just like the energy ground state. This could be an alley towards understanding how quantum behaviour could follow from a classical cellular automaton.

6. General relativity and black holes

Finally, there is General Relativity. This theory must be regarded as just an other theme of the general concept of local gauge theories. It represents a non compact gauge group of curved coordinate transformations and it may well be that it can be handled similarly. It is important to remember that this theory is not renormalizable when presented in its usual form. We do observe that the addition of one further interaction term, the square of the conformal Weyl curvature term $C_{\mu\nu\alpha\beta}$, restores renormalisability at the cost of negative energy modes [19]. Perhaps this mode can serve as a fast variable, but much more work will be needed to remedy various difficulties.

Theories for quantum mechanics that also aspire to include General Relativity, must address the fundamental black hole question. Black holes that are sufficiently large and heavy compared to the Planck scale of units, can be perfectly well described by classical, *i.e.* unquantised Einsteinian laws. However many researchers appear to arrive at the conclusion that there is something wrong with the black hole horizon, which might even involve the larger black holes. The origin of this suspicion is the emergence of 'firewalls' forming a curtain of destruction against particles entering (or leaving) the horizon. The firewalls originate from the Hawking particles that are expected to emerge in the more distant future.

The present author found that there exists a unique procedure to neutralise the firewalls, but it does not happen automatically. To see what may well happen, one should compute the effects that particles entering a black hole have on the Hawking particles leaving. It is not an act of destruction but a precisely calculable effect of repositioning those rays of out going material. The bottom line is that the *positions* of the out going particles are effected by the *momenta* of the in-going ones, and, because of quantum duality relating position to momentum, the same relation is found when going backwards in time: the momenta of the out-going particles are linked to the positions of the in-going ones.

These findings allow one to construct a unique expression for the *black hole evolution matrix*, only requiring very basic knowledge of the mathematics of GR and QM.

However, we also hit upon a more sobering difficulty, The region behind the horizon has to be used to describe the *time reverse* of the region normally visible, otherwise the evolution matrix (actually a quantum evolution matrix) fails completely to be unitary. For someone familiar with the Schwarzschild metric and its generalisations that have charge and angular momentum, there is no surprise here, but for the quantum physicist, this presents a problem. If we reverse the time direction, we also change the signs of all energies of the matter particles. Yet quantum field theories became successful precisely because they ensure the positivity of the energy of all particles. Can we allow ourselves a theory with such apparently conflicting properties?

The only answer that we could find is that we should act in a way similar to what P.A.M. Dirac did in order to overcome the negative energy problem in the Dirac equation. Now in a black hole, we have bosons and electrons alike, but we can achieve the same result by assuming that the entire band of energy eigenstates in a field theory should be bounded from below *and from above!* In that case, we can interpret the energy states beyond the horizon to be filled with particles completely, if the region at our side of the horizon is empty, and the other way around. The name *anti-vacuum* was coined, describing the completely filled state.

The region beyond the vacuum then represents a CPT inversion of the region at our side of the horizon. This picture appears to make perfectly sense, and we believe it to be likely that it resolves the energy inversion problem in black hole physics.

This solution of the energy inversion problem replaces the infinite energy spectrum of all harmonic oscillators generated by the fields outside the horizon, with a spectrum of evenly separated energy levels that have both a beginning and an end, the end being the highest possible energy level. We note that this is not only the energy spectrum of an atom with finite spin inside a homogeneous magnetic field (the Zeeman atom), but it also represents the energy levels of a periodic system with finite time steps δt in its evolution law, see the beginning of Section 2.

Indeed, we find that black holes may be telling us something about the origin of quantum mechanics.

7. Conclusions

Our aim was to rescue the concept of ontology as opposed to epistemology in quantum mechanics. This tells us that the atoms, molecules, electrons and other tiny entities are features of things that really exist. They evolve into different states or objects that also exist, according to universal physical laws. We find that this makes perfect sense if what we now perceive as quantum mechanics is understood as a vector representation of the states as they exist and evolve. Vector representations themselves allow superposition, and one finds that the superpositions of 'ontological' states evolve through the same Schrödinger equations as the original states themselves. This in turn implies that one may ignore everything that is said about ontological existence as long as we use Born's dictum that the absolute squares of the superposition coefficients represent probabilities. The reason why we nevertheless attach much importance to our ontological interpretation is that it implies a severe restriction for the evolution laws; asking for the existence of an ontological representation forces us to redesign the set of elementary basis elements of Hilbert space, which might implicate new constraints on what kinds of Standard Model we may suspect to describe our world.

An ontological interpretation is also of great help in resolving the numerous 'paradoxes' that have been around confusing scientists as well as young students as to what 'reality' really is about. Questions such as the physical process that seems to be associated to the 'collapse of the wave function', the 'measurement problem', as well as the difficulties raised in the EPR paper as well as Bell's theorem, questions surrounding the features of entanglement, and the Greenberger - Horne - Zeilinger (GHZ) paradox, all become much less counter intuitive and mysterious than what they look like in their original quantum settings.

The explanation of these features is that the real thing that is happening is the classically evolving collection of microscopic objects, of which the fastest periodically moving things automatically enter into a completely featureless, even distribution over all of their possible states.

Remarkably, the reason why the states of the fastest moving objects stay in an even distribution is better understood in the quantum formalism than when using the original classical picture: the highest energy excitations are difficult or almost impossible to excite, simply because the energy needed for that is usually unavailable to us: in our accelerators we can only reach a dozen or so TeVs, and in cosmic rays the highest detectable energies are still well below the Planck scale. Therefore, the excited modes are only virtually present, and may well be ignored in practice. And, since all superposition coefficients for the ground state are equal, the distribution is featureless, in practice – according to Born.

Thus, what we really find is that the lowest energy states of the slow variables become entangled due to their interactions with the fastest variables. Quantum mechanics ensues; it is mathematically inevitable.

Our work is far from finished. Fresh young minds should probe the remaining mysteries; in particular, the Standard Model is built from fundamental symmetry principles. There are more symmetries than one might have expected from 'just any' classical system: there are many continuous symmetries, and also non-compact symmetries such as Lorentz invariance and general coordinate transformation invariance, and there are exact local gauge invariances as dictated by the gauge fields in the Standard Model.

Finally, a natural place must be found where we can put and understand the black hole solutions of Einstein's equations. They too must obey the laws of quantum mechanics, before we can embrace these remarkable systems in our overall picture of nature. Data obtained from the observations of cosmologists must also be incorporated. What we are searching for is nothing less than a grand picture of the evolution laws shaping our physical world.

Author details

Gerard 't Hooft
Faculty of Science, Institute for Theoretical Physics, Utrecht University, Utrecht, The Netherlands

*Address all correspondence to: g.thooft@uu.nl

References

[1] A. Pais, *Inward bound: of matter and forces in the physical world,* Oxford University Press 1986.

[2] E. Schrödinger, *An Undulatory Theory of the Mechanics of Atoms and Molecules*, Phys. Rev. **28**, 1049 – Published 1 December 1926.

[3] M. Born, *Zur Quantenmechanik der Stoßvorgänge* (German) [Toward the quantum mechanics of collision processes], j-z-physik, **37** (1926) 863–867; id., j-z-physik, **38**(1026) 803–827.

[4] D. Bohm, *A Suggested Interpretation of the Quantum Theory in Terms of "Hidden Variables, I"*. Phys. Rev. **85** (1952) 166–179; id., *II*. Phys. Rev. **85** (1952) 180–193.

[5] Everett H. *Relative State Formulation of Quantum Mechanics*. Reviews of Mod. Phys. **29** (1957) 454

[6] B.S. DeWitt, in *The Many-Worlds Interpretation of Quantum Mechanics*, ed. B.S. DeWitt and N. Graham. Princeton University Press, 1973.

[7] van Kampen NG. *Ten theorems about quantum mechanical measurements*. Physica **A. 153** (1988) 97

[8] G,J. Sussman and J. Wisdom, *Numerical Evidence That the Motion of Pluto Is Chaotic*, Science, New Series, Vol. **241**, No. 4864. (Jul. 22, 1988), pp. 433 – 437.

[9] G. 't Hooft, *The Cellular Automaton Interpretation of Quantum Mechanics*, Fundamental Theories of Physics, Vol. 185, Springer International Publishing, 2016. eBook ISBN 978-3-319-41285-6, DOI 10.1007/978-3-319-41285-6; Hardcover ISBN 978-3-319-41284-9, Series ISSN 0168-1222, Edition Number 1 http://arxiv.org/abs/1405.1548.

[10] G. 't Hooft, *Deterministic Quantum Mechanics: the Mathematical Equations,* Subm. to the Article Collection: "Towards a Local Realist View of the Quantum Phenomenon", Frontiers Research Topic, ed. A. Casado et al, arxiv:2005.06374.

[11] G. 't Hooft, *Explicit construction of Local Hidden Variables for any quantum theory up to any desired accuracy* arxiv: 2103.04335[quant-ph], to be pub.

[12] von Neumann J. *Mathematical Foundations of Quantum Mechanics* (German edition 1932). Princeton: Princeton University Press.

[13] E. Nelson, *Derivation of the Schrödinger equation from Newtonian mechanics,* Phys. Rev. **150**, 1079-1085 (1966); E. Nelson, *Quantum Fluctuations,* Princeton University Press, Princeton, (1985).

[14] G. 't Hooft, *The Ontology Conservation Law as an Alternative to the Many World Interpretation of Quantum Mechanics,* presented at the Conference PAFT2019, on 'Current Problems in Theoretical Physics', session on 'Foundations of Quantum Theory', Vietri sul Mare, Salerno, Italy, April 13-17, 2019, arxiv:1904.12364; G. 't Hooft, *Constructing deterministic models for quantum mechanical systems*, Int. J. Geometric Methods in Modern Physics, Vol. **17**, No. supp01, 2040007 (2020)

[15] Bell JS. *On the Einstein Podolsky Rosen Paradox*. Physics. 1964;**1**:195

[16] J.F. Clauser, M.A. Horne, A. Shimony, and R.A. Holt, *Proposed experiment to test local hidden-variable theories*, Phys. Rev. Lett. **23** (15): 880–4, doi:10.1103 / Phys Rev Lett. 23.880 (1969).

[17] F. Jegerlehner, *The Standard Model of Particle Physics as a Conspiracy Theory and the possible Role of the Higgs Boson in*

the Evolution of the Early Universe, Acta Phys. Polonica B, special tribute to the Nobel Laureate Martinus Veltman, pp 575-605, Ed. M. Praszałowicz et al, 2021.

[18] G. 't Hooft, *The Big Questions in Elementary Particle Physics*, Acta Phys. Polonica B, special tribute to the Nobel Laureate Martinus Veltman, pp 841-854, Ed. M. Praszałowicz et al, 2021.

[19] G.'t Hooft, *Local conformal symmetry in black holes, standard model, and quantum gravity,* presented at the XIVth Marcel Grossmann Meeting, Rome, July 13, 2015, Int. J. Mod. Phys. **D 26**, No. 3 (2017) 1730006, DOI: 10.1142/S0218271817300063.

Chapter 3

Can We Entangle Entanglement?

Mrittunjoy Guha Majumdar

Abstract

In this chapter, nested multilevel entanglement is formulated and discussed in terms of *Matryoshka* states. The generation of such states that contain nested patterns of entanglement, based on an anisotropic XY model has been proposed. Two classes of multilevel-entanglement- the *Matryoshka Q-GHZ states* and *Matryoshka generalised GHZ states*, are studied. Potential applications of such resource states, such as for quantum teleportation of arbitrary one, two and three qubits states, bidirectional teleportation of arbitrary two qubit states and probabilistic circular controlled teleportation are proposed and discussed, in terms of a Matryoshka state over seven qubits. We also discuss fractal network protocols, surface codes and graph states as well as generation of arbitrary entangled states at remote locations in this chapter.

Keywords: Quantum Computation, Multipartite Entanglement, Quantum State Sharing

1. Introduction

Quantum Entanglement is a fundamental non-classical aspect of entities in the quantum realm, which disallows a reductionist description of a composite system - in terms of the state and properties of its quantum constituents. Erwin Schrodinger once famously said,

> *"Thus one disposes provisionally until the entanglement is resolved by actual observation of only a common description of the two in that space of higher dimension. This is the reason that knowledge of the individual systems can decline to the scantiest, even to zero, while that of the combined system remains continually maximal. The best possible knowledge of a whole does not include the best possible knowledge of its parts—and this is what keeps coming back to haunt us"*

Albert Einstein, Boris Podolsky and Nathan Rosen, famously known as EPR, and Schrödinger, who called it *Verschränkung*, highlighted the intrinsic order of statistical relations between the constituents of a compound quantum system, first recognised what they called a 'spooky' feature of the quantum world. John Bell showed that it is entanglement which irrevocably rules out the possibility of ascribing values to physical quantities of entangled systems prior to measurement. He accepted the EPR conclusion around the quantum description of nature not being 'complete', with the principles of 'realism' (measurement results are determined by properties that the particles carry prior to, and are independent of, the measurement), 'locality' (measurements obtained at one location are independent of any actions performed at another point that is spacelike separated) and 'free will' (settings of a local apparatus are independent of what EPR called 'hidden variables' that determine the local results) being primary in this discussion. Bell showed that if one were to assume these principles, then one obtains constraints in the form of certain inequalities, called Bell's Inequalities, on the statistical correlations in the measured values of properties of the systems, and that the probabilities of the outcomes of a measurement performed on constituents of an entangled system violate the Bell inequality. In this manner, it was shown that entanglement makes it impossible to simulate quantum correlations within the classical manner of thinking. Greenberger, Horne, and Zeilinger (GHZ) went beyond two particles in showing entanglement of quantum particles leads to contradictions with *Local Hidden Variables Models (LVHM)* for non-statistical predictions of quantum systems. During his doctoral studies at Université d'Orsay, Alain Aspect performed the first experimental realisation of the Bell's Inequalities.

Today, entanglement is instrumental in the formulation of information processing tasks in the quantum realm. It has been used in applications such as superdense coding and teleportation. *Bennett et al* first proposed a scheme for quantum teleportation, wherein a genuinely entangled Bell state was used to transmit an arbitrary single qubit [1]. Many different kinds of entangled quantum states have been used to teleport arbitrary quantum states since then, including Bell states [2, 3], GHZ states [4, 5], W states [6, 7] and multiqubit states [8–10]. There have been hop-by-hop and multi-hop quantum teleportation schemes proposed since then as well as schemes to teleport GHZ-like states using two types of four-qubit states [11, 12]. Teleportation has been proposed in two-copy quantum teleportation scheme [13], using cluster states [14], in higher dimensions [15] and also shown to be possible over atmospheric channels [16]. More recently, various derivatives of the standard teleportation scheme have been proposed, including those used for bidirectional teleportation [15, 17, 18], controlled teleportation [19, 20], quantum operation sharing [21, 22], quantum secret sharing [23–25] and arbitrated quantum teleportation [26, 27]. For multiple participants in a quantum information processing task, entangled multiqubit states and multipartite entanglement play the preeminent role, with multiqubit resource states varying from GHZ- and W-states to clusters states [28]. Lately, W-GHZ composite states have been used for remote state preparation, teleportation and superdense coding of arbitrary quantum states [29, 30]. *Shuai et al* showed how GHZ-GHZ channels can be used for bidirectional quantum communication [31]. The physical realisation of such composite systems have been explored in a number of physical platforms such as using cavity QED [32]. Properties of spin squeezing when multi-qubit GHZ state and W state are superposed have also been studied [33]. These composite quantum states contain varying degrees of multilevel and genuine multipartite entanglement, which can be used for applications in quantum information processing [34, 35]. *Yang et al* investigated the feasibility of experimentally creating GHZ states comprising of three logical qubits in a decoherence-free subspace, by using superconducting transmon qutrits coupled to a co-planar waveguide resonator [36].

Since not all forms of entanglement are relevant for distinct information processing applications, the determination of resource states for specific information processing tasks is of paramount importance. This, along with any characteristic protection or resilience against noise and decoherence provided by a resource state, forms the underlying principle of quantum resource theories [37–40]. In the latter pursuit, decoherence-free subspaces provide a natural solution and associated resources to produce quantum resource-states that are not easily decohered [41–44]. Stabiliser codes are a resource that constitutes a crucial ingredient for effective quantum error correction [45], while cluster states are resource states that are used for measurement-based quantum computation and error corrections [46–51]. Certain realisations of a standard resource-state have more resilience against decoherence, such as in the case of cluster states generated with Ising-type interactions, wherein the entanglement in the state persisted upto a fairly large number of measurements on the qubits to disentangle them [52]. These resource state display various distinct forms of entanglement: some are maximally entangled, such as resource-states used for teleportation, while others are partially entangled, such as in the case of cluster states. In the case of cluster states, the partial entanglement is a resource in itself, since the one requires a specific protection of the 'quantumness' and correlations in the segments of the state against perturbations or measurements of other segments of the state. If the resource-state were maximally entangled, such a measurement or perturbation of one segment will collapse the state of the remaining segments to a specific state, thereby not maintaining the system as a viable quantum resource for further cluster operations. If we were to generalise and extend this idea to conceptualise states that maintain near maximal entanglement in segments of the state while maintaining weak correlations between the segments, we could have interesting resource-states and associated applications of such states. This is the central idea and motivation behind generalising the concept of *Matryoshka* states: *Matryoshka Generalised GHZ* states, *Matryoshka GHZ-Bell* States and *Matryoshka Q-GHZ* States.

In multi-qubit quantum states, an important property is that entanglement is monogamous - quantum entanglement cannot be shared freely among various parties. *Osborne and Verstraete* showed that the entanglement for bipartitions over an n-qubit system follows a monogamy relation [53]:

$$\tau\left(\rho_{A_1A_2}\right)+\tau\left(\rho_{A_1A_3}\right)+\ldots+\tau\left(\rho_{A_1A_n}\right) \leq \tau\left(\rho_{A_1(A_2\ldots A_n)}\right) \tag{1}$$

where $\tau\left(\rho_{A_1(A_2\ldots A_n)}\right)$ denotes the bipartite quantum entanglement measured by the tangle across the bipartition $A_1 : A_2A_3\ldots A_n$. In this chapter, we discuss the weak coupling between near-maximally entangled (sub)states due to the constraint placed by entanglement monogamy [54–57]. The concept of Matryoshka states was first given by *Di Franco et al* [58], with the name 'Matryoshka' coming from the Russian word for 'nesting doll'. The underlying concept of a Matryoshka state is genuine entanglement in multilevel systems, with the entanglement in higher level systems being more than or equal to the entanglement in the lower level constituents:

$$\mathcal{E}_{d_i} \geq \mathcal{E}_{d_j}, d_i > d_j \tag{2}$$

where E_{d_i} is the entanglement measure of the level d_i. In this chapter, we will discuss the characteristics and applications of two classes of Matryoshka states for $d = 2$ multiqubit systems, which are as follows:

1. *Matryoshka Generalised GHZ states*

$$|\psi_{MGHz}\rangle = \sum_{k=1}^{L} \lambda_k |GHZ_{d_1}^{a_{k,d_1},\pm}\rangle \ldots |GHZ_{d_N}^{a_{k,d_N},\pm}\rangle \tag{3}$$

$$\left\langle GHZ_{d_i}^{a_{k,d_i},\pm} | GHZ_{d_i}^{a_{k',d_i},\pm}\right\rangle = \delta_{kk'} \forall i \tag{4}$$

A particular case of such states are the *Matryoshka GHZ-Bell states*

$$|\psi_{MGHzB}\rangle = \sum_{k=1}^{L} \lambda_k |GHZ_{d_1}^{a_{k,d_1},\pm}\rangle |B_{d_2}^{a_{k,d_2},\pm}\rangle \ldots |B_{d_N}^{a_{k,d_N},\pm}\rangle \tag{5}$$

where $|B\rangle$ signifies a Bell state.

$$\left\langle GHZ_{d_1}^{a_{k,d_1},\pm} | GHZ_{d_1}^{a_{k',d_1},\pm} \right\rangle = \delta_{kk'} \forall i \tag{6}$$

$$\left\langle B_{d_i}^{a_{k,d_i},\pm} | B_{d_i}^{a_{k',d_i},\pm} \right\rangle = \delta_{kk'} \forall i \tag{7}$$

2.*Matryoshka Q-GHZ states*

$$|\psi_{MExG}\rangle = \sum_{k=1}^{L} \lambda_k |A_1^k\rangle |GHZ_{d_2}^{a_{k,d_2},\pm}\rangle \dots |GHZ_{d_N}^{a_{k,d_N},\pm}\rangle \tag{8}$$

$$\left\langle GHZ_{d_i}^{a_{k,d_i},\pm} | GHZ_{d_i}^{a_{k',d_i},\pm} \right\rangle = \delta_{kk'} \forall i, \left\langle A_1^k | A_1^{k'} \right\rangle = \delta_{kk'} \tag{9}$$

where $|A\rangle$ are orthogonal states that are eigenstates in the Z-basis for all qubits in the state. Here the subscript 'd_i' in $|GHZ_{d_i}^{a_{k,d_i}}\rangle$ denotes the number of qubits in the ith subsystem, while a is the decimal representation of the superposed term in the GHZ-like state that has the lowest decimal representation and $\pm$ denotes the relative phase between the terms in superposition. GHZ-like states are the states that can be created from the GHZ state using local unitary operations. So, for instance, in a three-qubit system $|GHZ^{2,+}\rangle = \frac{1}{\sqrt{2}}(|010\rangle + |101\rangle)$ can be created from $|GHZ\rangle = \frac{1}{\sqrt{2}}(|000\rangle + |111\rangle)$ using $I_{2\times2} \otimes \sigma_x \otimes I_{2\times2}$, or in other words - we apply a qubit flip σ_x operation on the second qubit, leaving the other qubits untouched. In the summation above, $L = 2^{n_h}$ where n_h is the number of qubits in the largest subsystem.

Nomenclature and Acronyms Used. GHZ state is a multipartite maximally entangled state, first defined for three qubits: $|\psi_\pm\rangle = \frac{1}{\sqrt{2}}(|000\rangle \pm |111\rangle)$. A Hadamard Operator is a quantum logical gate that acts on a single qubit and maps the basis state $|0\rangle$ to $\frac{|0\rangle+|1\rangle}{\sqrt{2}}$ and $|1\rangle$ to $\frac{|0\rangle-|1\rangle}{\sqrt{2}}$.

2. Localised correlation generation: how can we generate entangled entanglement?

Matryoshka states can be generated in various physical platforms, such as in spin systems and in trapped ions. Fröwis and Dür [59] studied the stability of superpositions of macroscopically distinct quantum states under decoherence, wherein they looked at realising concatenated-GHZ states: $|\phi_C\rangle = \frac{1}{\sqrt{2}}\left(|GHZ_m^+\rangle^{\otimes N} + |GHZ_m^-\rangle^{\otimes N}\right)$ (with $|GHZ_N^\pm\rangle = \frac{1}{\sqrt{2}}\left(|0\rangle^{\otimes N} \pm |1\rangle^{\otimes N}\right)$), which is a Matryoshka generalised state state, in trapped ion systems. The underlying principle to realise entangled entanglement is to have localised and intra-level correlation generation, which begins with creation of entanglement in one level, thereafter entanglement of this entangled structure over higher-level basis states and so on. For the purposes of this chapter, we will be considering the GHZ and GHZ-like states as the primary unit of entanglement.

The algorithm for generating entangled entanglement in a system comprising of GHZ and GHZ-like states as the units of entanglement is given by

Step 1: Creation of a ground state $|0000\dots0\rangle$ with total number of qubits being $n = 3k$ for some finite, non-vanishing integer k.

Step 2: Application of a Hadamard gate on the $(3n+1)^{\text{th}}$ qubits to give $|+00+\dots0\rangle$ where $|+\rangle = \frac{1}{\sqrt{2}}(|0\rangle + |1\rangle)$.

Step 3: Application of *CNOT* operation with the $(3n+1)^{\text{th}}$ qubits as the control for the corresponding $(3n+2)^{\text{th}}$ qubits and $(3n+3)^{\text{th}}$ qubits as target to give a state of the form $|GHZ^{(123)}\rangle|GHZ^{(456)}\rangle \ldots |GHZ^{(n-2,n-1,n)}\rangle$.

Step 4: Application of composite operation of the form of $\sum_{i=0}^{n/3} P_{3i+1}P_{3i+2}P_{3i+3}$ where P represents Pauli operations or combination of Pauli operations such as $\sigma_x\sigma_z$ and

1. $P_{3i+1} \neq P_{3j+1}$ for $i \neq j \forall i,j \in \mathbb{Z}$

2. $P_{3i+2} \neq P_{3j+2}$ for $i \neq j \forall i,j \in \mathbb{Z}$

3. $P_{3i+3} \neq P_{3j+3}$ for $i \neq j \forall i,j \in \mathbb{Z}$

2.1 Generation of Matryoshka states using spin systems in condensed matter physics

In this chapter, the generation of Matryoshka states will be explored in spin systems in condensed matter physics. Unlike in the case of the aforementioned algorithm, instead of composite operators, in this case we have localised generation and minimal interactions between different GHZ and GHZ-like states to create the Matryoshka states. In this case, we consider N spin-$\frac{1}{2}$ particles, with each spin coupled to its nearest neighbours by the XY Hamiltonian

$$H = \sum_{i=1}^{N-1} (J_{X,i}\hat{X}_i\hat{X}_{i+1} + J_{Y,i}\hat{Y}_i\hat{Y}_{i+1}) \tag{10}$$

where $J_{\sigma,i}$ is the pairwise coupling constant with $\sigma = \hat{X}, \hat{Y}, \hat{Z}$ being the Pauli operators. For the purposes of this chapter, we take N to be odd. *Franco et al* [58] showed that it is sufficient to state that the information flux between the $\hat{X}$ ($\hat{Y}$) operators of the first and last qubits in the spin-chain depends on an alternating set of coupling strengths. For example, the information flux from $\hat{X}_1$ to $\hat{X}_N$ depends only on the set $\{J_{Y,1}, J_{X,2}, \ldots, J_{Y,N-1}\}$ and is independent of any other coupling rate in the spin-chain. *Christandl et al* [60, 61] showed that after a time $t^* = \pi/\lambda$ with λ being a scaling constant (as mentioned in the definition of the case of a perfect state transfer in a linear spin-chain given by weighted coupling strengths: $J_{\sigma,i} = \lambda\sqrt{i(N-i)}$), the state of the first qubit in the spin-chain can be perfectly transferred to the last qubit. We see that by preparing the initial state of this spin-chain in an completely separable eigenstate of the tensorial product of Z_i operators, say $|\Psi(0)\rangle = |000\ldots0\rangle_{12\ldots N}$, we obtain an information flux towards symmetric two-site spin operators, and a final state of the form [58].

$$|\psi_0\rangle = |0\rangle_c \otimes_{i=0}^{M} |\psi_+\rangle_{2i+1,N-2i} \otimes_{i=1}^{M} |\psi_-\rangle_{2i,N-2i+1} \tag{11}$$

$$|\psi_1\rangle = |1\rangle_c \otimes_{i=0}^{M} |\psi_-\rangle_{2i+1,N-2i} \otimes_{i=1}^{M} |\psi_+\rangle_{2i,N-2i+1} \tag{12}$$

where c labels the central site of the spin-chain, $M = \frac{N-3}{4}$ and $|\psi_\pm\rangle = \frac{1}{\sqrt{2}}(|00\rangle \pm |11\rangle)$. An illustration of the setup has been shown in **Figure 1**.

The critical step in the creation of the Matryoshka GHZ-Bell state is the evolution of the central and two neighbouring qubits to the GHZ state, without disturbing the rest of the spin-chain. This is a key result around the generation of *Matryoshka* GHZ-Bell states in this chapter, which can be extended to other classes of *Matryoshka* states.

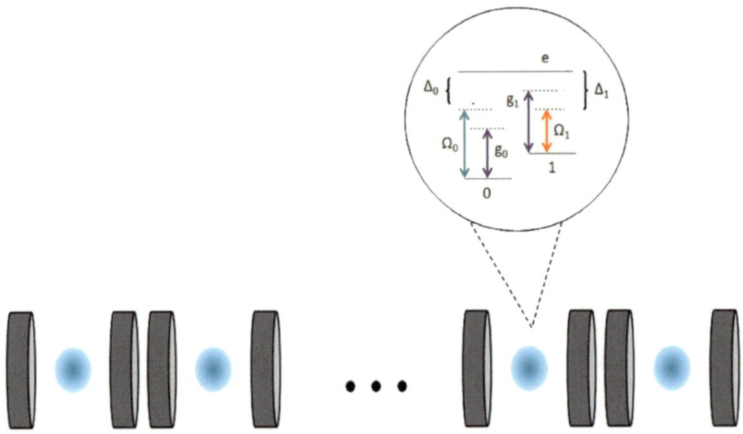

Figure 1.
Scheme for the generation of Matryoshka GHZ-Bell resource-states, where the effective spin–spin XY Hamiltonianan is obtained as an effective adiabatic Hamiltonian for a linear chain of optical cavities with each interacting with a three-level atomic system. The ground states of each atomic unit provide the computational space of each spin, and the dipole-forbidden transition between these states is realised as an (adiabatic) Raman transition through the excited state: $|e\rangle_i$ with $i = 1, 2, \ldots, N$. The cavity field drives off-resonantly the dipole-allowed channel $|j\rangle_i \leftrightarrow |e\rangle_i$ with the Rabi frequency g_j, $j = 0, 1$. Two lasers are also coupled to these atomic transitions with strength Ω_j and detuning Λ_j.

For this, we need to switch off all the interactions except for those connecting the central qubit to the neighbouring ones. A point to note here is that had we started with $|\Psi(0)\rangle = |111...1\rangle_{12...N}$, we would have obtained a final state of the form

$$|\psi_0\rangle = |0\rangle_c \otimes_{i=0}^{M} |\psi_-\rangle_{2i+1,N-2i} \otimes_{i=1}^{M} |\psi_+\rangle_{2i,N-2i+1} \tag{13}$$

$$|\psi_1\rangle = |1\rangle_c \otimes_{i=0}^{M} |\psi_+\rangle_{2i+1,N-2i} \otimes_{i=1}^{M} |\psi_-\rangle_{2i,N-2i+1} \tag{14}$$

We use this principle and the idea that after evolution over time t^*, the states in Eqs. (2) and (3) transform back to $|000...000\rangle_{12...N}$ and states in Eqs. (4) and (5) transform back to $|111...11\rangle_{12...N}$. We can utilise this concept, by taking the state in Eq. (2) and evolving it, for the truncated subsystem comprising of the central qubit and the adjoining qubits. A point to note here is that due to only coupling that connects to the central qubits, the coupling strength ($J'_{\sigma,i} = \lambda' \sqrt{i(3-i)}$) and time of evolution ($t'' = \pi/\lambda'$) vary accordingly. Before carrying out this evolution, we perform a Hadamard operation on the central qubit to give

$$|\psi_0\rangle = \frac{1}{\sqrt{2}}(|0\rangle_c + |1\rangle_c) \otimes_{i=0}^{M} |\psi_+\rangle_{2i+1,N-2i} \otimes_{i=1}^{M} |\psi_-\rangle_{2i,N-2i+1} \tag{15}$$

We now perform the truncated subsystem time-evolution with the parameters (J', t'') to give us the state

$$|\psi_0\rangle = \frac{1}{\sqrt{2}}(|000\rangle + |111\rangle)_{c-1,c,c+1} \otimes_{i=0}^{M-1} |\psi_+\rangle_{2i+1,N-2i} \otimes_{i=1}^{M} |\psi_-\rangle_{2i,N-2i+1} \tag{16}$$

Therefore, we can obtain a Matryoshka GHZ-Bell state using nearest spin–spin interactions in a spin-chain. A similar generation protocol can be defined for the other two classes of Matryoshka states. The teleportation of an arbitrary n-qubit state can be performed using Matryoshka GHZ-Bell States [62].

Given the triangular three-qubit configuration, we can also consider the anisotropic Heisenberg Hamiltonian, which describes the interaction between three spins that are located at the corners of an equilateral triangle lying in the xy-plane, as shown in **Figure 2**.

$$H = -J_{xy}\sum_{i=1}^{3}\left(S_i^x S_{i+1}^x + S_i^y S_{i+1}^y\right) - J_z\sum_{i=1}^{3} S_i^z S_{i+1}^z + H_z \tag{17}$$

here the three spins S_i, with S = 1/2, are located at the corners i = 1, 2, 3, and $S_1 = S_4$. J_{xy} and J_z are the in-plane and out-of-plane exchange coupling constants respectively, and $H_Z = \sum_{i=1}^{3} b_i.S_i$ denotes the Zeeman coupling of the spins S_i to the externally applied magnetic fields b_i at the sites i. If we consider isotropic exchange couplings: $J_{xy} = J_z = J > 0$ (ferromagnetic coupling) and $b_i = 0 \forall i$, we have a ground-state qudruplet that is spanned by the GHZ states: $\frac{1}{\sqrt{2}}(|000\rangle + |111\rangle)$ and $\frac{1}{\sqrt{2}}(|000\rangle - |111\rangle)$, along with the W- and spin-flipped W-states. A set of appropriately chosen magnetics fields will allow us to split off an approximate GHZ state from this degenerate eigenspace. If we find a set of magnetic fields that, in classical spin systems, shall result in exactly two degenerate minima for the configurations $|000\rangle$, representing the $\downarrow\downarrow\downarrow$ spin configuration, and $|111\rangle$, representing the $\uparrow\uparrow\uparrow$ spin configuration, with an energy barrier in between, quantum mechanical tunnelling shall yield the desired states. The magnetic fields must be of the same strength, in-plane and sum to zero, with a convenient additional choice being that of the field pointing radially outward. Therefore, the successive directions of the magnetic fields have to differ by an angle of $2\pi/3$ with respect to each other. Going by the schematic in **Figure 2**, we can write the hamiltonian

$$\begin{aligned} H = {} & -J_{xy}\sum_{i=1}^{3}\left(S_i^x S_{i+1}^x + S_i^y S_{i+1}^y\right) - J_z\sum_{i=1}^{3} S_i^z S_{i+1}^z + H_z \\ & + \sum_{il=1}^{N_l}\left[-J_{xy}^{(il)}\sum_{i=1}^{3}\left(S_i^x S_{i+1}^x + S_i^y S_{i+1}^y\right) - J_z^{(il)}\sum_{i=1}^{3} S_i^z S_{i+1}^z + H_z^{(il)}\right] \\ & \sum_{ir=1}^{N_r}\left[-J_{xy}^{(ir)}\sum_{i=1}^{3}\left(S_i^x S_{i+1}^x + S_i^y S_{i+1}^y\right) - J_z^{(ir)}\sum_{i=1}^{3} S_i^z S_{i+1}^z + H_z^{(ir)}\right] \\ & + \sum_{il=1}^{N_l}\lambda_{(il,il+1)}^{l} S_i^{n_r}.S_{i+1}^{n_l} + \sum_{ir=0}^{N_r-1}\lambda_{(ir,ir+1)}^{r} S_i^{n_r}.S_{i+1}^{n_l} \end{aligned} \tag{18}$$

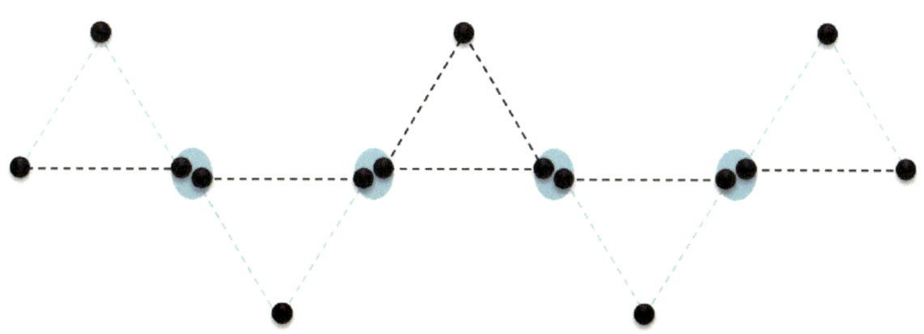

Figure 2.
Schematic for all (three) classes of Matryoshka states for d = 2 levels of the quantum system, explored in this chapter. The triangular formations encapsulate the logical units of two/three qubits mediated by CNOT gates. Each of these triangular units are weakly coupled to each other (shown with light blue patches). In the case of the Matryoshka GHZ-Bell *states, we only have the black links, while for the* Matryoshka Generalised GHZ *states and* Matryoshka Q-GHZ *states, we also have the blue links.*

where the superscripts il and ir denote the left and right branches respectively of the schematic arounnd a central triangular unit. For $il = 1$, we have the leftmost triangular unit and for $ir = N_r$, we have the rightmost triangular unit. N_l and N_r denote the number of units on the left and right side of the central triangular unit. In principle, we can have an asymmetric case where $N_l \neq N_r$. In the fourth line, the term S_{N_l+1} and S_0 refer to the spins in the central triangular unit connected to the adjacent left and right triangular units respectively. Moreover, both $\lambda^l_{(il,il+1)}$ and $\lambda^l_{(ir,ir+1)}$ are coupling constants between adjacent triangular units that are numerically negligible with respect to J but are non-zero, to account for inter-unit coupling. $S_i^{n_r}$ and $S_i^{n_l}$ are right and left connecting nodes of the ith triangular unit.

An important point here is the condition: $\left\langle GHZ_{d_i}^{a_k,d_i,\pm} | GHZ_{d_i}^{a_{k'},d_i,\pm} \right\rangle = \delta_{kk'} \forall i, \left\langle A_1^k | A_1^{k'} \right\rangle = \delta_{kk'}$ in Eqs. (4), (6) and (9). This is ensured by the additional application of single qubit gates on the nodes of the triangular units. For instance, $\frac{1}{\sqrt{2}}(|000\rangle + |111\rangle) \xrightarrow{\sigma_x^2} \frac{1}{\sqrt{2}}(|010\rangle + |101\rangle$. Using combination of such single qubit operations, we can span the entire space of GHZ and GHZ-like states. The important point here is the synchronised timing of these operations, with the inter-unit coupling, so as to give us a superposition over orthogonal GHZ and GHZ-like states for all triangular units, as shown in **Figure 2**.

3. Creating tesselated networks of Matryoshka states

The Matryoshka Generalised GHZ states can also be oriented in a tesselated manner, as shown in **Figure 3(a)** for the case of symmetric 3-qubit GHZ triangular units. The Matryoshka GHZ-Bell states, a specific form of these states, can even be oriented in an emanatory manner, as shown in **Figure 3(b)**. These two orientations can be used for tessellation in three-dimensions, as in the case of the spherical configuration shown in **Figure 3(c)**, which shows the method of lattice surgery (discussed later in the chapter). More complex forms such as the hexagonal-pentagonal tiling with 6-qubit and 5-qubit GHZ states can be used for forms such as truncated icosahedrons. Lastly, we can also have higher GHZ-forms in a self-similar, fractal manner, as shown in **Figure 3(d)**. Each of these configurations will be studied in the *Application* section of this chapter. An interesting future direction of pursuing this line of research would be in squeezed baths, which *Zippilli et al* studied and showed that a squeezed bath, which acts on the central element of a harmonic chain, could drive the entire system to a steady state that features a series of nested entangled pairs of oscillators [63]. This series ideally covers the entire chain regardless of its size. Extending this result to higher number of nearest neighbour interactions is non-trivial.

4. Where can we use entangled entanglement?

Matryoshka states have a second level of entanglement (nesting) and have additional protection against loss of coherence under local transformations.

4.1 Fractal network protocol

In this chapter, a new quantum communication architecture is being proposed, whereby there are levels of entanglement which underly a distributed network. If we have

DOI: http://dx.doi.org/10.5772/intechopen.98535

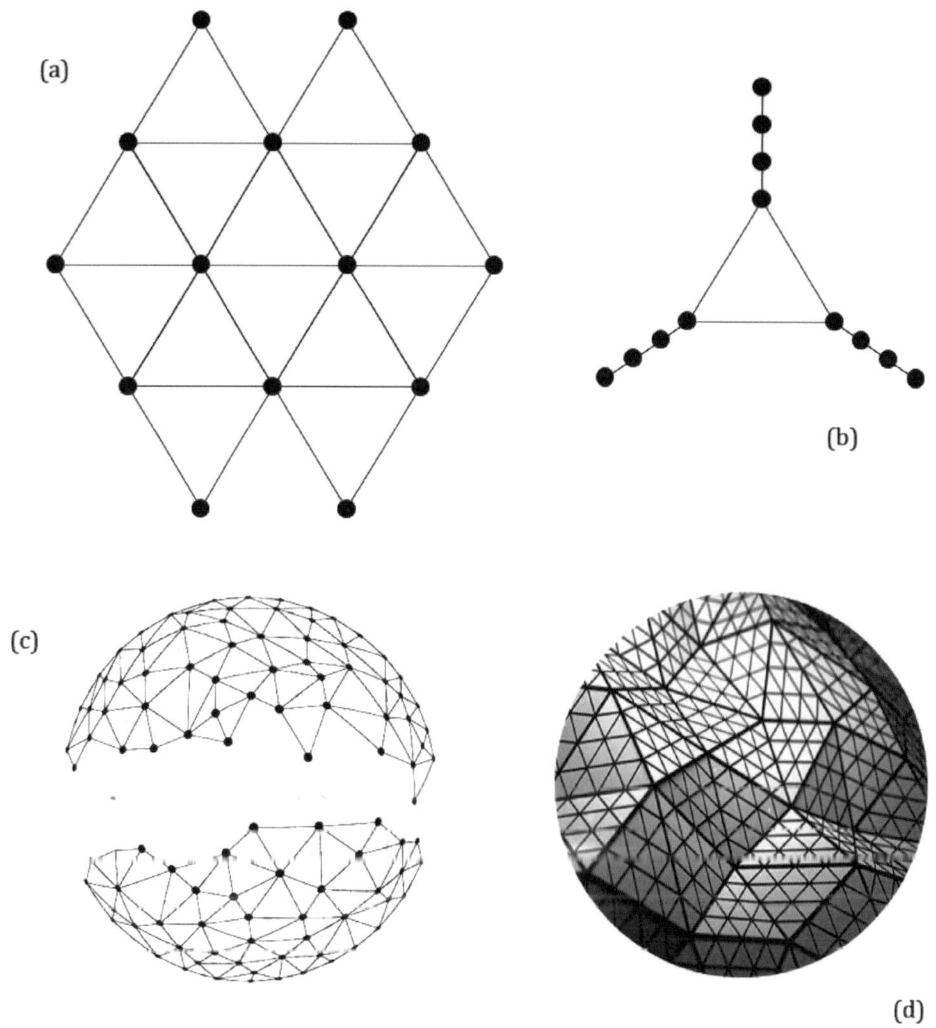

Figure 3.
The various tesselation patterns possible with the GHZ triangular units in (a) generalised GHZ states in a planar tesselated format, (b) GHZ-Bell states with an emanatory geometry, (c) spherical pattern created by planar codes, along with illustration of lattice surgery with projective measurements, and (d) hierarchical GHZ-state levels, where we have a self-similar nature of the tesselation. A point to note here is that each node in the diagram has three physical qubits (one from each GHZ triangular unit) in the generalised GHZ states and two physical qubits in the GHZ-Bell states.

$$|0\rangle_L^n = \frac{1}{\sqrt{2}}\left(|0_L^{n-1}0_L^{n-1}0_L^{n-1}\rangle + |1_L^{n-1}1_L^{n-1}1_L^{n-1}\rangle\right) \tag{19}$$

$$|1\rangle_L^n = \frac{1}{\sqrt{2}}\left(|0_L^{n-1}0_L^{n-1}0_L^{n-1}\rangle - |1_L^{n-1}1_L^{n-1}1_L^{n-1}\rangle\right) \tag{20}$$

As you can see, these are special cases of Matryoshka Generalised GHZ states, with the superscript n defining the layer of the network. A point to note here is that $n = 1$ is the layer with physical qubits, and so $|0\rangle_L^1 = |0\rangle$ and $|1\rangle_L^1 = |1\rangle$. This effectively creates layers of *entangled entanglement*. This is highly useful in providing multiple levels of protection in quantum network encoding. The key point here is the heralded nature in which we can access levels from the highest to the lowest, with a projective measurement onto the basis logical qubits of the just-lower level of entanglement to pass through a level of entanglement-enabled security and robustness.

4.2 Surface codes, graph states and cluster states

We can define effective surface codes with Matryoshka states, with triangular units. The primary operation proposed to be utilised in this regard is that of lattice surgery and merging. Topological encoding of quantum data facilitates information processing to be protected from the effects of decoherence on physical qubits, by having a logical qubit encoded in the entangled state of many physical qubits. Among the various codes used for this purpose, the surface code has the highest tolerance of component error, when implemented on a two-dimensional lattice of spin-qubits with nearest-neighbour interactions [64–68]. Mhalla and Perdrix [69] proved that the application of measurements in the (X, Z) plane, with one-qubit measurement as per the basis

$$\{ \cos\theta|0\rangle + \sin\theta|1\rangle, \sin\theta|0\rangle - \cos\theta|1\rangle\} \tag{21}$$

for some θ over graph states that are represented by triangular grids, is a universal model of quantum computation. A point to note here is that, for any θ, the observable associated with the measurement in this basis is $\cos 2\theta Z + \sin 2\theta X$. For a given simple undirected graph $G = (V, E)$ of order n, where V represent vertices and E edges, the graph state $|G\rangle$ is the unique quantum state such that for any vertex $u \in V$,

$$X_u Z_{\mathcal{N}(u)}|G\rangle = |G\rangle \tag{22}$$

The Pauli operators constitute a group acting on a set V of n qubits is generated by $X_u, Z_u, i.I_{u\in V}$, where I is the identity, X_u and Z_u are operators that act as identity on the neighbourhood of u and with the following action on vertex u

$$X : |x\rangle \rightarrow |\overline{x}\rangle \tag{23}$$

$$Z : |x\rangle \rightarrow (-1)^x|\overline{x}\rangle \tag{24}$$

In our circuit, we will have to project three physical qubits from three adjacent triangular units to a single subspace for implementing this model. If we consider the state: $\frac{1}{2\sqrt{2}}(|00_c0\rangle+|11_c1\rangle)(|00_c0\rangle+|11_c1\rangle)(|00_c0\rangle+|11_c1\rangle)$, with the subscript c denoting the physical qubits adjacent to each other and that are projected to a single subspace. If we initialise an ancilla qubit in the state $|+\rangle = \frac{1}{\sqrt{2}}(|0\rangle+|1\rangle)$ and use the conditional rotation gate

$$U_\gamma = \begin{pmatrix} 1 & 0 & 0 & 0 \\ 0 & 1 & 0 & 0 \\ 0 & 0 & \cos\frac{\gamma}{2} & \sin\frac{\gamma}{2} \\ 0 & 0 & -\sin\frac{\gamma}{2} & \cos\frac{\gamma}{2} \end{pmatrix} \tag{25}$$

and apply this sequentially with the three adjacent physical qubits (with subscript 'c') and the ancilla as target, we project the ancilla to a unique state that can be retained for the graph state that is thereby defined, by going over the entire tessellated lattice of triangular GHZ-units.

4.3 Establishing multiparticle entanglement between nodes of a quantum communication network

We can use the unique form of the asymmetric Matryoshka Generalised GHZ states to establish multipartite entanglement between nodes of a quantum communication network. The important part about this protocol is the role of projection measurements on a central terminal. Considering a Matryoshka GHZ-Bell state with an m-particle GHZ state and n-terminals in a quantum network

$$|\psi_{MGHzB}\rangle = \sum_{k=1}^{L} \lambda_k |GHZ_m^{a_{k,m},\pm}\rangle |B_{d_1}^{a_{k,d_1},\pm}\rangle \dots |B_{d_n}^{a_{k,d_n},\pm}\rangle \tag{26}$$

where $|B\rangle$ signifies a Bell state, $\left\langle GHZ_m^{a_{k,m},\pm} | GHZ_m^{a_{k',m},\pm} \right\rangle = \delta_{kk'} \forall i$ and $\left\langle B_{d_i}^{a_{k,d_i},\pm} | B_{d_i}^{a_{k',d_i},\pm} \right\rangle = \delta_{kk'} \forall i$. Each user has one particle of a Bell-state, while the other particle of the Bell-state is with the central terminal. Measuring the particles of the Bell-pairs at the central terminal in a basis defined by maximally entangled states over n-qubits will project the distant qubits into maximally n-qubit entangled states as well. In fact, it need not only be one n-qubit maximally entangled state at the spatially distant nodes but could be multiple (partially or maximally) entangled states of varying number of qubits connecting different permutations of end-terminals, depending on the projective measurement performed on the central terminal. Some examples of such remote establishment of entanglement have been shown in **Figure 4**.

4.4 Quantum networks, repeater protocols and quantum communication

Quantum networks can facilitate the realisation of quantum technologies such as distributed quantum computing [70], secure communication schemes [71] and quantum metrology [72–75]. In our formalism for GHZ-based network protocols, the key element is that of being able to merge GHZ triangular units, which is done by projecting states at adjacent nodes into a single subspace (as shown in **Figure 5**), as has been tried on atomic systems previously [76]. A *generalised GHZ-GHZ Matryoshka state* can also assist in the recovery of quantum network operability upon node failure, based on the formalism given by *Guha Majumdar and Srinivas Garani* [77].

4.5 Teleportation and superdense coding

Let us look at the applications of such nested entanglement with the example of a state close to a Matryoshka Q-GHZ state: the Xin-Wei Zha (XZW) State. *Xin-Wei Zha et al* [78] discovered a genuinely entangled seven-qubit state through a numerical optimization process, following the path taken by *Brown et al* [79] and *Borras et al* [80] to find genuinely entangled five-qubit and six-qubit states:

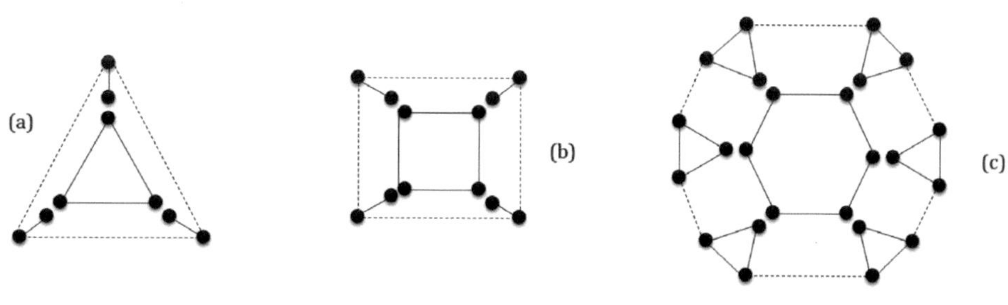

Figure 4.
Illustration of networks for entanglement generation in remote nodes in (a) triangular format (b) rectangular format (c) polyhedra (dodecagon) format, with distinct patterns of entanglement generated at the periphery depending on the projective measurements at the central terminal(s).

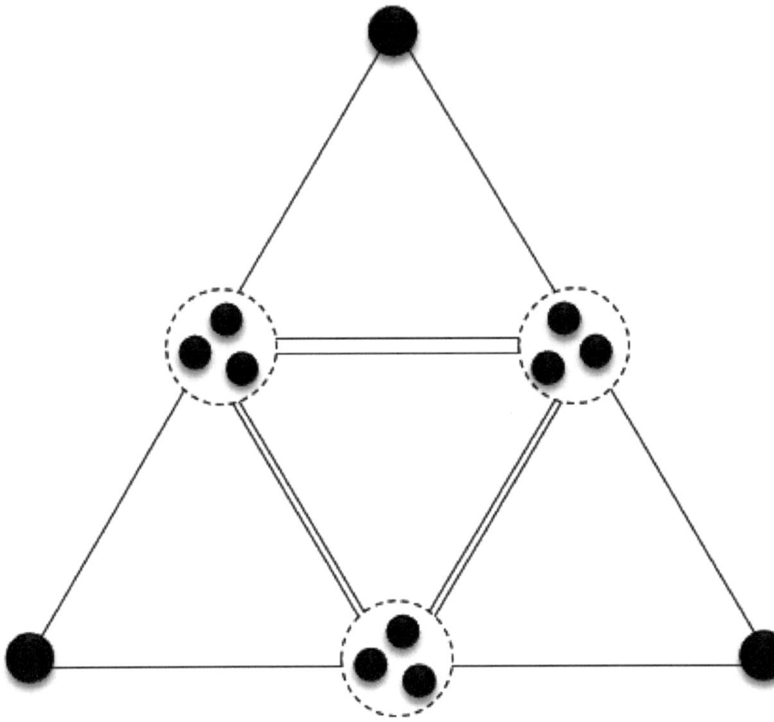

Figure 5.
Network repeater protocol with three-qubit projective measurements at nodes to create higher-distance entangled networks.

$$|\psi_7\rangle = \frac{1}{2\sqrt{2}}(|000\rangle_{135}|\psi_+\rangle_{24}|\psi_+\rangle_{67} + |001\rangle_{135}|\phi_-\rangle_{24}|\phi_+\rangle_{67}$$
$$+|010\rangle_{135}|\psi_-\rangle_{24}|\phi_-\rangle_{67} + |011\rangle_{135}|\phi_+\rangle_{24}|\psi_-\rangle_{67} \quad (27)$$
$$+|100\rangle_{135}|\phi_+\rangle_{24}|\phi_+\rangle_{67} + |101\rangle_{135}|\psi_-\rangle_{24}|\psi_+\rangle_{67}$$
$$+|110\rangle_{135}|\phi_-\rangle_{24}|\psi_-\rangle_{67} + |111\rangle_{135}|\psi_+\rangle_{24}|\phi_-\rangle_{67})$$

This state is a specific form of the Q-GHZ State defined in Eq. (6), with $\lambda_k \forall k = \frac{1}{2\sqrt{2}}$ and $|A_1^k \in \{|000\rangle, |001\rangle, |010\rangle, |011\rangle, |100\rangle, |101\rangle, |110\rangle, |111\rangle\}$. Another point to note here is that the GHZ states here are for $d = 2$, thereby effectively being the Bell states. This resource state can be used for teleportation of arbitrary single, double and triple qubit states. The 3 (Q State)-2 (Bell State)-2 (Bell State) structure of the resource-state, given in Eq. (17), helps us in devising a quantum circuit to generate the state, as shown in **Figure 6** and realised on *IBM Quantum Experience*. To obtain the resource-state, we apply a unitary operator on qubits 1, 3 and 5: $U = I_{4\times4} \oplus (\sigma_z \otimes \sigma_z)$.

This state has marginal density matrices for subsystems over one or two qubits that are completely mixed, with $\pi_{ij} = Tr_{ij}\rho_{ij}^2 = \frac{1}{4}\ \forall\ i,j \in \{1,2,3,4,5,6,7\}, i<j$, $\pi_i = Tr_i\rho_i^2 = \frac{1}{2}\ \forall\ i \in \{1,2,3,4,5,6,7\}$. For three-qubit subsystems, some of the partitions have mixed marginal density matrices: $\pi_{ijk} = Tr_{ijk}\rho_{ijk}^2 = \frac{1}{8}\ \forall\ i,j \in \{1,2,3,4,5,6,7\}$, $i<j<k \wedge (ijk) \neq (127), (367), (457)$ and $\pi_{127} = \pi_{367} = \pi_{457} = \frac{1}{4}$.

The seven-qubit genuinely entangled resource state $|\Gamma_7\rangle$ can be used for a number of applications, such as quantum secret sharing (*Supplementary Material* A.1, A.2 and A.3), the perfect linear teleportation of an arbitrary one-qubit state (*Supplementary Material* B.1.1), probabilistic circular teleportation of arbitrary one-

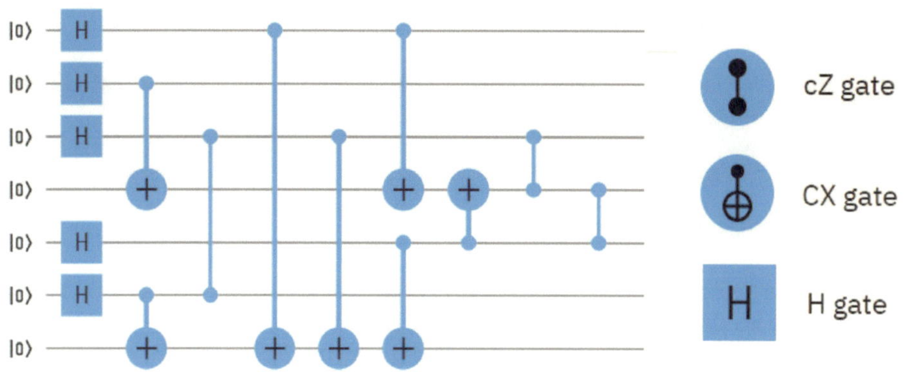

Figure 6.
Quantum circuit for the generation of the seven-qubit genuinely entangled state, on IBM Quantum Experience. *Here* CX gate *is the CNOT gate,* cZ gate *is the CPHASE gate and* H gate *is the Hadamard gate.*

qubit states (*Supplementary Material* B.1.2), perfect linear teleportation of an arbitrary two-qubit state (*Supplementary Material* B.2.1), bidirectional teleportation of arbitrary two-qubit states (*Supplementary Material* B.2.2) and perfect linear teleportation of an arbitrary three-qubit state (*Supplementary Material* B.3).

5. Conclusion

In this chapter, the generation and application of nested entanglement in Matryoshka resource-states for quantum information processing was studied. A novel scheme for the generation of such quantum states has been proposed using an anisotropic XY spin–spin interaction-based model. The application of the *Matryoshka GHZ-Bell states* for n-qubit teleportation is reviewed and an extension of this formalism to more general classes of *Matryoshka states* is posited. An example of a state close to a perfect *Matryoshka Q-GHZ state* is given in the form of the genuinely entangled seven-qubit *Xin-Wei Zha state*. Generation, characterisation and application of this seven-qubit resource state is presented. This work should lay the groundwork for other studies into the area of nested entanglement, including forays into higher layers of nesting entanglement. Particularly, the problem of composite quantum states containing nested entanglement can be explored further, theoretically and experimentally, be it in surface codes, establishment of multipartite entanglement in quantum networks, teleportation, superdense coding and more broadly in quantum communication protocols. The main advantage of the model and method presented in this chapter is the accessibility of the condensed matter system presented, while the primary limitation of the model presented in this chapter is the need for fine-tuning of various interaction terms that have to be time-sequenced very carefully. The concept of entangled entanglement is the key result of the chapter, which can be implemented with other non-trivial combination of unitary transformations over multiple qubits.

Acknowledgements

I would like to acknowledge the guidance and contribution of Prof. Prasanta Panigrahi, IISER-Kolkata. This work was supported by the Homi Bhabha Centre for Science Education, Tata Institute of Fundamental Research, Mumbai, India.

Data availability statement

Data sharing is not applicable to this article as no new data were created or analysed in this study.

A. Quantum Secret Sharing

Quantum Secret Sharing (QSS) is a procedure for splitting a message into several parts so that no single subset of parts is sufficient to read the message, but the entire set is. This can also naturally be extended to Quantum Operation Sharing (QOS). In this section, quantum secret sharing using the 7 qubit XZW resource-state is proposed, with three proposals for the same.

A.1 Proposal 1

Let us consider the situation in which Alice possesses the 1st qubit, Bob possesses qubits 2, 3, 4, 5, 6 and Charlie possesses the 7th qubit. Alice has an unknown qubit $\alpha|0\rangle + \beta|1\rangle$ which she wants to share with Bob and Charlie.

Now, Alice combines the unknown qubit with $|\Psi_7\rangle$ and performs a Bell measurement, and conveys her outcome to Charlie by two classical bits. For instance if Alice measures in the $|\Phi_+\rangle$ basis, then the Bob-Charlie system evolves into the entangled state.

$$\begin{aligned}
&\alpha|100001\rangle - \alpha|000100\rangle - \alpha|000111\rangle - \alpha|001001\rangle \\
&+\alpha|001010\rangle + \alpha|010101\rangle - \alpha|010110\rangle - \alpha|011000\rangle \\
&+\alpha|011011\rangle + \alpha|100010\rangle + \alpha|101100\rangle + \alpha|101111\rangle - \\
&\alpha|110011\rangle + \alpha|111101\rangle - \alpha|111110\rangle + \beta|000000\rangle \\
&+\beta|000011\rangle + \beta|001101\rangle + \beta|001110\rangle + \beta|010001\rangle \\
&-\beta|010010\rangle + \beta|011100\rangle - \beta|011111\rangle - \beta|100101\rangle - \\
&\beta|000000\rangle - \beta|100110\rangle + \beta|101000\rangle + \beta|101011\rangle \\
&+\beta|110100\rangle - \beta|110111\rangle - \beta|111001\rangle + \beta|111010\rangle
\end{aligned} \tag{A1}$$

Now, Bob can perform a five-qubit measurement and convey his outcome to Charlie through a classical channel. Having known the outcome of both their measurement, Charlie will obtain a certain single qubit quantum state. The outcome of the measurement performed by Bob is correlated with the state obtained by Charlie. If Bob measures $|A_{\pm}\rangle$ then Charlie obtains the state $\alpha|0\rangle \pm \beta|1\rangle$, while if Bob measures the state $|B_{\pm}\rangle$ then Charlie obtains the state $\beta|0\rangle \pm \alpha|1\rangle$, where

$$\begin{aligned}
&|A_{\pm}\rangle \\
&= -|00010\rangle + |00101\rangle - |01011\rangle - |01100\rangle \\
&+|10001\rangle + |10110\rangle - |11111\rangle \pm (|00001\rangle \\
&+|00110\rangle + |01000\rangle - |01111\rangle - |10010\rangle \\
&+|10101\rangle - |11011\rangle - |11100\rangle)
\end{aligned} \tag{A2}$$

$$
\begin{aligned}
&|B\pm\rangle \\
&\quad = \pm(|10000\rangle - |00011\rangle - |00100\rangle + |01010\rangle \\
&\quad +|01101\rangle + |10111\rangle - |11001\rangle + |11110\rangle) \\
&\quad +|00000\rangle + |00111\rangle - |01001\rangle + |01110\rangle \\
&\quad -|00000\rangle - |10011\rangle + |10100\rangle + |11010\rangle \\
&\quad +|11101\rangle
\end{aligned}
\tag{A3}
$$

A.2 Proposal 2

Let us consider the situation in which Alice possesses the qubits 1 and 2, Bob possesses qubits 3, 4, 5 and 6 and Charlie possesses the 7th qubit. Alice has an unknown qubit $\alpha|0\rangle + \beta|1\rangle$ which she wants to share with Bob and Charlie. Now Alice can measure in a particular basis. Suppose she measures in the GHZ Basis. Now, Bob can perform a four-qubit measurement and convey his outcome to Charlie through a classical channel. Having known the outcome of both their measurement, Charlie will obtain a certain single qubit quantum state. The outcome of the measurement performed by Bob and the state obtained by Charlie is given as follows: if Bob measures states $|x_{\pm}\rangle$, Charlie obtains states $\alpha|0\rangle \pm \beta|1\rangle$, while if Bob measures states $|Y_{\pm}\rangle$ then Charlie obtains the states $\beta|0\rangle \pm \alpha|1\rangle$, where $|x_{\pm}\rangle = \frac{1}{4}\alpha|0000\rangle + \alpha|0111\rangle + \alpha|1001\rangle + \alpha|1110\rangle \pm (\beta|1001\rangle + \beta|0000\rangle + \beta|1110\rangle - \beta|0111\rangle)$ and $|Y_{\pm}\rangle = \frac{1}{4}\alpha|0001\rangle + \alpha|0110\rangle + \alpha|1000\rangle - \alpha|1111\rangle \pm (\beta|1000\rangle + \beta|0001\rangle - \beta|1111\rangle - \beta|0110\rangle)$

A.3 Proposal 3

Let us consider the situation in which Alice possesses the qubits 1, 2, 3 and 4, Bob possesses qubits 5 and 6 and Charlie possesses the 7th qubit. Alice has an unknown qubit $\alpha|0\rangle + \beta|1\rangle$ which she wants to share with Bob and Charlie. Based on the state Alice measures $(|A_i\rangle \forall i \in \{1,2,3,4,5,6,7,8\})$, Bob and Charlie obtain a corresponding state $|BC_i\rangle$, where

$$|A_1\rangle = \frac{1}{4}(|01111\rangle - |01011\rangle + |10010\rangle + |11001\rangle + |11100\rangle + |11101\rangle - |11000\rangle)$$

$$|A_2\rangle = \frac{1}{4}(|01111\rangle + |01011\rangle - |10010\rangle - |11001\rangle - |11100\rangle + |11101\rangle - |11000\rangle)$$

$$|A_3\rangle = \frac{1}{4}(|01111\rangle + |01011\rangle + |10010\rangle + |11001\rangle + |11100\rangle - |11101\rangle + |11000\rangle)$$

$$|A_4\rangle = \frac{1}{4}(|01111\rangle - |01011\rangle - |10010\rangle - |11001\rangle - |11100\rangle - |11101\rangle + |11000\rangle)$$

$$|A_5\rangle = \frac{1}{4}(|11111\rangle - |11011\rangle + |00010\rangle + |01001\rangle + |01100\rangle + |01101\rangle - |01000\rangle)$$

$$|A_6\rangle = \frac{1}{4}(|11111\rangle + |11011\rangle - |00010\rangle - |01001\rangle - |01100\rangle + |01101\rangle - |01000\rangle)$$

$$|A_7\rangle = \frac{1}{4}(|11111\rangle + |11011\rangle + |00010\rangle + |01001\rangle + |01100\rangle - |01101\rangle + |01000\rangle)$$

$$|A_8\rangle = \frac{1}{4}(|11111\rangle - |11011\rangle - |00010\rangle - |01001\rangle - |01100\rangle - |01101\rangle + |01000\rangle)$$

and

$$
\begin{aligned}
|BC_1\rangle &= \alpha|1\rangle|\Phi_-\rangle + \alpha|0\rangle|\Psi_-\rangle + \beta|0\rangle|\Phi_+\rangle + \beta|1\rangle|\Psi_+\rangle \\
|BC_2\rangle &= \alpha|1\rangle|\Phi_-\rangle - \alpha|0\rangle|\Psi_-\rangle - \beta|0\rangle|\Phi_+\rangle + \beta|1\rangle|\Psi_+\rangle \\
|BC_3\rangle &= \alpha|1\rangle|\Phi_-\rangle - \alpha|0\rangle|\Psi_-\rangle + \beta|0\rangle|\Phi_+\rangle - \beta|1\rangle|\Psi_+\rangle \\
|BC_4\rangle &= \alpha|1\rangle|\Phi_-\rangle + \alpha|0\rangle|\Psi_-\rangle - \beta|0\rangle|\Phi_+\rangle - \beta|1\rangle|\Psi_+\rangle \\
|BC_5\rangle &= \beta|1\rangle|\Phi_-\rangle + \beta|0\rangle|\Psi_-\rangle + \alpha|0\rangle|\Phi_+\rangle + \alpha|1\rangle|\Psi_+\rangle \\
|BC_6\rangle &= \beta|1\rangle|\Phi_-\rangle - \beta|0\rangle|\Psi_-\rangle - \alpha|0\rangle|\Phi_+\rangle + \alpha|1\rangle|\Psi_+\rangle \\
|BC_7\rangle &= \beta|1\rangle|\Phi_-\rangle - \beta|0\rangle|\Psi_-\rangle + \alpha|0\rangle|\Phi_+\rangle - \alpha|1\rangle|\Psi_+\rangle \\
|BC_8\rangle &= \beta|1\rangle|\Phi_-\rangle + \beta|0\rangle|\Psi_-\rangle - \alpha|0\rangle|\Phi_+\rangle - \alpha|1\rangle|\Psi_+\rangle
\end{aligned}
$$

Bob can now perform a Bell measurement on his particles, and Charlie can obtain a particular resultant state by applying the appropriate unitary operation.

For example, if the joint-state obtained by Bob and Charlie is $\beta|1\rangle|\Phi_-\rangle + \beta|0\rangle|\Psi_-\rangle - \alpha|0\rangle|\Phi_+\rangle - \alpha|1\rangle|\Psi_+\rangle$, one can see that Charlie will obtain the state $|C_i\rangle,\ i = 1, 2, 3, 4$ corresponding to the state measured by Bob $|B_i\rangle$, where $|B_1\rangle = \frac{1}{\sqrt{2}}|01\rangle,\ |B_2\rangle = \frac{1}{\sqrt{2}}|10\rangle,\ |B_3\rangle = \frac{1}{\sqrt{2}}|11\rangle,\ |B_4\rangle = \frac{1}{\sqrt{2}}|00\rangle$ and $|C_1\rangle = \alpha|0\rangle + \beta|1\rangle,\ |C_2\rangle = \alpha|0\rangle - \beta|1\rangle,\ |C_3\rangle = \alpha|1\rangle + \beta|0\rangle,\ |C_4\rangle = \alpha|1\rangle - \beta|0\rangle$

B. Quantum teleportation

B.1 Quantum teleportation of arbitrary one-qubit state

B.1.1 Linear teleportation scheme

To begin with, an arbitrary single qubit state can be teleported using the resource state $|\Gamma_7\rangle$ will be considered. In this case Alice possesses qubits 1, 2, 3, 4, 5, 6 and the 7th particle belongs to Bob. Alice wants to transport an arbitrary state $|\psi^{(1)}\rangle = \alpha|0\rangle + \beta|1\rangle$ to Bob. The combined state of the system is $|\Gamma_7^{(1)}\rangle = |\psi^{(1)}\rangle \otimes |\Gamma_7\rangle$. Alice measures the seven qubits in her possession via the seven qubit orthonormal states:

$$
\begin{aligned}
|\xi^{\pm}\rangle = &|0000\rangle|\Psi^0_{GHZ}\rangle - |0001\rangle|\Psi^3_{GHZ}\rangle + |0010\rangle|\Psi^7_{GHZ}\rangle \\
&+|0011\rangle|\Psi^4_{GHZ}\rangle - |0100\rangle|\Psi^5_{GHZ}\rangle - |0101\rangle|\Psi^6_{GHZ}\rangle \\
&+|0110\rangle|\Psi^2_{GHZ}\rangle + |0111\rangle|\Psi^1_{GHZ}\rangle \pm (|1000\rangle|\Psi^2_{GHZ}\rangle \\
&-|1001\rangle|\Psi^1_{GHZ}\rangle - |1010\rangle|\Psi^5_{GHZ}\rangle - |1011\rangle|\Psi^6_{GHZ}\rangle \\
&+|1100\rangle|\Psi^7_{GHZ}\rangle + |1101\rangle|\Psi^4_{GHZ}\rangle + |1110\rangle|\Psi^0_{GHZ}\rangle \\
&-|1111\rangle|\Psi^3_{GHZ}\rangle)
\end{aligned} \tag{A4}
$$

$$
\begin{aligned}
|\nu^{\pm}\rangle = &|1000\rangle|\Psi^0_{GHZ}\rangle - |1001\rangle|\Psi^3_{GHZ}\rangle + |1010\rangle|\Psi^7_{GHZ}\rangle \\
&+|1011\rangle|\Psi^4_{GHZ}\rangle - |1100\rangle|\Psi^5_{GHZ}\rangle - |1101\rangle|\Psi^6_{GHZ}\rangle \\
&+|1110\rangle|\Psi^2_{GHZ}\rangle + |1111\rangle|\Psi^1_{GHZ}\rangle \pm (|0000\rangle|\Psi^2_{GHZ}\rangle \\
&-|0001\rangle|\Psi^1_{GHZ}\rangle - |0010\rangle|\Psi^5_{GHZ}\rangle - |0011\rangle|\Psi^6_{GHZ}\rangle \\
&+|0100\rangle|\Psi^7_{GHZ}\rangle + |0101\rangle|\Psi^4_{GHZ}\rangle + |0110\rangle|\Psi^0_{GHZ}\rangle \\
&-|0111\rangle|\Psi^3_{GHZ}\rangle)
\end{aligned} \tag{A5}
$$

where $|\Psi^{0,1}_{GHZ}\rangle = \frac{1}{\sqrt{2}}[|000\rangle \pm |111\rangle]$, $|\Psi^{2,3}_{GHZ}\rangle = \frac{1}{\sqrt{2}}[|001\rangle \pm |110\rangle]$, $|\Psi^{4,5}_{GHZ}\rangle = \frac{1}{\sqrt{2}}[|010\rangle \pm |101\rangle]$ and $|\Psi^{6,7}_{GHZ}\rangle = \frac{1}{\sqrt{2}}[|100\rangle \pm |011\rangle]$.

Alice then conveys the outcome of the measurement results to Bob via two classical bits. Bob then applies a suitable unitary operation from the set $I, \sigma_x, i\sigma_y, \sigma_z$ to recover the original state, sent by Alice. In this way, one can teleport an arbitrary single-qubit state using the state $|\Gamma_7\rangle$.

B.1.2 Probabilistic circular teleportation scheme for arbitrary one-qubit states

Not only is the seven-qubit resource state useful for linear and bidirectional teleportation but can also facilitate the probabilistic teleportation of an arbitrary single-qubit states in a circular manner between three network-nodes (users). Let us say we have Alice, Bob and Charlie in the system, with the first qubit used as a control qubit, qubits 1 and 4 given to Alice, qubits 2 and 6 given to Bob and qubits 3 and 7 given to Charlie. Let us say the arbitrary states are $|\psi_A\rangle = \alpha_A|0_A\rangle + \beta_A|1_A\rangle$, $|\psi_B\rangle = \alpha_B|0_B\rangle + \beta_B|1_B\rangle$ and $|\psi_C\rangle = \alpha_C|0_C\rangle + \beta_C|1_C\rangle$. Then, the composite state is given by $|\psi_A\rangle \otimes |\psi_B\rangle \otimes |\psi_C\rangle \otimes |\Gamma_7\rangle_{TA_1B_1C_1A_1B_1C_1}$, where $|\Gamma_7\rangle_T$ is the control qubit. We apply a CNOT gate using the qubits A, B and C of the arbitrary states as the control-qubits and the first qubits of each user as the target-qubit. Let us for simplicity only consider the case where $|\Gamma_7\rangle_T = |0\rangle$.

Let us now measure the first qubits of Alice, Bob and Charlie in the Z-basis. Let us say $|\Gamma_7\rangle_{A_1B_1C_1} = |010\rangle$, then we have the state $|\psi''\rangle = -\frac{1}{4}((|001\rangle\ |010\rangle)\alpha_A\alpha_B\alpha_C|0_A0_B0_C\rangle + (|100\rangle\ |111\rangle)\alpha_A\alpha_B\beta_C|0_A0_B1_C\rangle + (|000\rangle + |011\rangle)\alpha_A\beta_B\alpha_C|0_A1_B0_C\rangle + (|101\rangle + |110\rangle)\alpha_A\beta_B\beta_C|0_A1_B1_C\rangle + (|100\rangle\ |111\rangle)\beta_A\alpha_B\alpha_C|1_A0_B0_C\rangle + (-|001\rangle + |010\rangle)\beta_A\alpha_B\beta_C|1_A0_B1_C\rangle + (-|101\rangle\ |110\rangle)\beta_A\beta_B\alpha_C|1_A1_B0_C\rangle + (|000\rangle + |011\rangle)\beta_A\beta_B\beta_C|1_A1_B1_C\rangle)$. We can now measure the control qubits in the X-basis. So, let us say, we have $|Q_AQ_BQ_C\rangle = |+_{A-B+C}\rangle$, then we obtain the state $|C_1\rangle(|A_1\rangle(-|B_1\rangle + \chi|B_3\rangle - \chi^{-1}|B_4\rangle - |B_2\rangle) + |A_2\rangle(-|B_1\rangle + \chi|B_3\rangle + \chi^{-1}|B_4\rangle + |B_2\rangle)) + |C_2\rangle(|A_1\rangle(-|B_1\rangle - \chi|B_3\rangle - \chi^{-1}|B_4\rangle + |B_2\rangle) + |A_2\rangle(-|B_1\rangle - \chi|B_3\rangle + \chi^{-1}|B_4\rangle - |B_2\rangle)) + |C_3\rangle(|A_1\rangle(|B_4\rangle - \chi|B_2\rangle - \chi^{-1}|_B\rangle - |B_3\rangle) + |A_2\rangle(|B_4\rangle - \chi|B_2\rangle + \chi^{-1}|B_1\rangle - B_3\rangle)) + |C_4\rangle(|A_1\rangle(|B_4\rangle + \chi|B_2\rangle - \chi^{-1}|B_1\rangle - |B_3\rangle) + |A_2\rangle(|B_4\rangle + \chi|B_2\rangle + \chi^{-1}|B_1\rangle + B_3\rangle))$, where $|C_1\rangle = \beta_C|0\rangle + \alpha_C|1\rangle$, $|C_2\rangle = \beta_C|0\rangle - \alpha_C|1\rangle$, $|C_3\rangle = \beta_C|1\rangle + \alpha_C|0\rangle$, $|C_4\rangle = \beta_C|1\rangle - \alpha_C|0\rangle$, $|B_1\rangle = \alpha_B|1\rangle + \beta_B|0\rangle$, $|B_2\rangle = \alpha_B|1\rangle - \beta_B|0\rangle$, $|B_3\rangle = \alpha_B|0\rangle + \beta_B|1\rangle$, $|B_4\rangle = \alpha_B|0\rangle - \beta_B|1\rangle$, $\chi = \frac{a_2}{a_1}$ with $a_1 = \beta_A + \alpha_A$, $a_2 = \alpha_A - \beta_A$, $|A_1\rangle = a_1|0\rangle + a_2|1\rangle$, $|A_1\rangle = a_1|0\rangle - a_2|1\rangle$. Therefore I see that the users can obtain states derived from the original state of the users next to them (Alice $\rightarrow$ Bob $\rightarrow$ Charlie $\rightarrow$ Alice). However, as you can see, this can be done in a probabilistic manner with one of the users not quite obtaining the original state but rather a derivative-state based on the original.

B.2 Quantum teleportation of arbitrary two-qubit state

B.2.1 Linear teleportation scheme

Similarly, an arbitrary two qubit quantum state can be teleported using the resource-state. In this case Alice possesses qubits 1, 2, 3, 4 and 5, and the 6th and 7th particles belong to Bob. Alice wants to transport an arbitrary state $|\psi^{(2)}\rangle = \alpha|00\rangle + \mu|10\rangle + \gamma|01\rangle + \beta|11\rangle$ to Bob. The combined state of the system is $|\Gamma_7^{(2)}\rangle = |\psi^{(2)}\rangle \otimes |\Gamma_7\rangle$,

$$
\begin{aligned}
|\Gamma_7^{(2)}\rangle = {} & \alpha(A_{00}|00\rangle + A_{01}|01\rangle + A_{10}|10\rangle + A_{11}|11\rangle) \\
& + \mu(B_{00}|00\rangle + B_{01}|01\rangle + B_{10}|10\rangle + B_{11}|11\rangle) \\
& + \gamma(C_{00}|00\rangle + C_{01}|01\rangle + C_{10}|10\rangle + C_{11}|11\rangle) \\
& + \beta(D_{00}|00\rangle + D_{01}|01\rangle + D_{10}|10\rangle + D_{11}|11\rangle)
\end{aligned} \tag{A6}
$$

where

$$
\begin{aligned}
A_{00} &= |0000\rangle|\Psi^0_{GHZ}\rangle + |0001\rangle|\Psi^4_{GHZ}\rangle - |0010\rangle|\Psi^5_{GHZ}\rangle - |0011\rangle|\Psi^7_{GHZ}\rangle, \\
A_{11} &= |0000\rangle|\Psi^1_{GHZ}\rangle + |0001\rangle|\Psi^5_{GHZ}\rangle|0010\rangle|\Psi^2_{GHZ}\rangle|0011\rangle|\Psi^7_{GHZ}\rangle, \\
A_{01} &= |0000\rangle|\Psi^6_{GHZ}\rangle - |0001\rangle|\Psi^2_{GHZ}\rangle + |0010\rangle|\Psi^4_{GHZ}\rangle + |0011\rangle|\Psi^0_{GHZ}\rangle, \\
A_{10} &= -|0000\rangle|\Psi^7_{GHZ}\rangle - |0001\rangle|\Psi^3_{GHZ}\rangle + |0010\rangle|\Psi^5_{GHZ}\rangle + |0011\rangle|\Psi^1_{GHZ}\rangle, \\
B_{00} &= |1000\rangle|\Psi^0_{GHZ}\rangle + |1001\rangle|\Psi^4_{GHZ}\rangle|1010\rangle|\Psi^2_{GHZ}\rangle + |1011\rangle|\Psi^6_{GHZ}\rangle, \\
B_{11} &= |1000\rangle|\Psi^0_{GHZ}\rangle + |1001\rangle|\Psi^5_{GHZ}\rangle - |1010\rangle|\Psi^3_{GHZ}\rangle - |1011\rangle|\Psi^7_{GHZ}\rangle, \\
B_{01} &= |1000\rangle|\Psi^6_{GHZ}\rangle - |1001\rangle|\Psi^2_{GHZ}\rangle + |1010\rangle|\Psi^4_{GHZ}\rangle + |1011\rangle|\Psi^0_{GHZ}\rangle, \\
B_{10} &= -|1000\rangle|\Psi^7_{GHZ}\rangle - |1001\rangle|\Psi^3_{GHZ}\rangle + |1010\rangle|\Psi^5_{GHZ}\rangle + |1011\rangle|\Psi^1_{GHZ}\rangle, \\
C_{00} &= |0100\rangle|\Psi^0_{GHZ}\rangle + |0101\rangle|\Psi^4_{GHZ}\rangle - |0110\rangle|\Psi^3_{GHZ}\rangle + |0111\rangle|\Psi^6_{GHZ}\rangle, \\
C_{11} &= |0100\rangle|\Psi^1_{GHZ}\rangle + |0101\rangle|\Psi^5_{GHZ}\rangle - |0110\rangle|\Psi^3_{GHZ}\rangle - |0111\rangle|\Psi^7_{GHZ}\rangle, \\
C_{01} &= |0100\rangle|\Psi^6_{GHZ}\rangle - |0101\rangle|\Psi^2_{GHZ}\rangle|0110\rangle|\Psi^4_{GHZ}\rangle + |0111\rangle|\Psi^0_{GHZ}\rangle, \\
C_{10} &= |0100\rangle|\Psi^7_{GHZ}\rangle - |0101\rangle|\Psi^3_{GHZ}\rangle + |0110\rangle|\Psi^5_{GHZ}\rangle + |0111\rangle|\Psi^1_{GHZ}\rangle, \\
D_{00} &= |1100\rangle|\Psi^0_{GHZ}\rangle + |1101\rangle|\Psi^4_{GHZ}\rangle - |1110\rangle|\Psi^2_{GHZ}\rangle + |1111\rangle|\Psi^6_{GHZ}\rangle, \\
D_{11} &= |1100\rangle|\Psi^1_{GHZ}\rangle + |1101\rangle|\Psi^5_{GHZ}\rangle|1110\rangle|\Psi^3_{GHZ}\rangle|1111\rangle|\Psi^7_{GHZ}\rangle, \\
D_{01} &= |1100\rangle|\Psi^6_{GHZ}\rangle - |1101\rangle|\Psi^2_{GHZ}\rangle + |1110\rangle|\Psi^4_{GHZ}\rangle + |1111\rangle|\Psi^0_{GHZ}\rangle, \\
D_{10} &= |1100\rangle|\Psi^6_{GHZ}\rangle - |1101\rangle|\Psi^3_{GHZ}\rangle + |1110\rangle|\Psi^5_{GHZ}\rangle + |1111\rangle|\Psi^1_{GHZ}\rangle,
\end{aligned}
$$

where $|\Psi^{0,1}_{GHZ}\rangle = \frac{1}{\sqrt{2}}[|000\rangle \pm |111\rangle]$, $|\Psi^{2,3}_{GHZ}\rangle = \frac{1}{\sqrt{2}}[|001\rangle \pm |110\rangle]$, $|\Psi^{4,5}_{GHZ}\rangle = \frac{1}{\sqrt{2}}[|010\rangle \pm |101\rangle]$ and $|\Psi^{6,7}_{GHZ}\rangle = \frac{1}{\sqrt{2}}[|100\rangle \pm |011\rangle]$. Now, Bob can carry out a combination of unitary operations, according to the given table, to obtain the original state teleported by Alice.

State Obtained by Alice	**Unitary Operation by Bob**
$A_{01} + B_{11} + C_{00} + D_{01}$	$I \otimes \sigma_x$
$A_{01} + B_{11} - C_{00} - D_{01}$	$\sigma_z \otimes \sigma_x$
$A_{01} - B_{11} + C_{00} - D_{01}$	$I \otimes i\sigma_y$
$A_{01} - B_{11} - C_{00} + D_{01}$	$\sigma_z \otimes i\sigma_y$
$A_{11} + B_{01} + C_{10} + D_{00}$	$\sigma_x \otimes \sigma_x$
$A_{11} - B_{01} + C_{10} - D_{00}$	$\sigma_x \otimes i\sigma_y$
$A_{11} + B_{01} - C_{10} - D_{00}$	$i\sigma_y \otimes \sigma_x$

State Obtained by Alice	Unitary Operation by Bob
$A_{11} - B_{01} - C_{10} + D_{00}$	$i\sigma_y \otimes i\sigma_y$
$A_{00} + B_{10} + C_{01} + D_{11}$	$I \otimes I$
$A_{00} - B_{10} + C_{01} - D_{11}$	$I \otimes \sigma_z$
$A_{00} + B_{10} - C_{01} - D_{11}$	$\sigma_z \otimes I$
$A_{00} - B_{10} - C_{01} + D_{11}$	$\sigma_z \otimes \sigma_z$
$A_{10} + B_{11} + C_{00} + D_{01}$	$\sigma_x \otimes I$
$A_{10} - B_{11} + C_{00} - D_{01}$	$\sigma_x \otimes \sigma_z$
$A_{10} + B_{11} - C_{00} - D_{01}$	$i\sigma_y \otimes I$
$A_{10} - B_{11} - C_{00} + D_{01}$	$i\sigma_y \otimes \sigma_z$

B.2.2 Bidirectional teleportation of arbitrary two-qubit states

The resource-state can also be used for bidirectional quantum teleportation. Bidirectional Controlled Quantum Teleportation (BCQT) protocols have been proposed for multi-qubit resource states, such as five-qubit [81], six-qubit [82, 83], seven-qubit [84–86] and eight-qubit states [87]. Bidirectional Controlled Quantum Teleportation can teleport arbitrary states between two users under the supervision of a third party. Zha et al proposed the first scheme for BCQT of single qubit states using a maximally entangled seve-qubit quantum state [85]. There have been schemes proposed for BCQT that utilise states with the same number of qubits as the quantum channel being used, and thereby realise bidirectional teleportation of arbitrary single-and two-qubit states under the controller Charlie [84, 86].

Let us say Alice and Bob would like to teleport two-qubit states to each other by utilizing the seven-qubit genuinely entangled resource state. We assume the form of the two-qubit states to be

$$|\phi\rangle_{A_1A_2} = \alpha_0|00\rangle + \alpha_1|01\rangle + \alpha_2|10\rangle + \alpha_3|11\rangle \quad \text{(A7)}$$

$$|\phi\rangle_{B_1B_2} = \beta_0|00\rangle + \beta_1|01\rangle + \beta_2|10\rangle + \beta_3|11\rangle \quad \text{(A8)}$$

For the resource-state, let Alice have the qubits 1,4 and 7, while Bob has the qubits 2, 3 and 6 and Charlie has the qubit 5. The steps for the scheme are as follows:

- Alice measures qubit 7 of the resource state and A_1 in the bell basis.
- Bob measures qubit 2 of the resource state and B_1 in the bell basis.
- Charlie, Alice and Bob measure their qubits in the Z basis.
- Alice and Bob measure their qubits A_2 and B_2 in the X-basis.
- We apply unitary transformations to the composite state to now get Alice's initial arbitrary state in Bob's terminal and Bob's initial arbitrary state in Alice's terminal.

We will now be looking more closely at these steps with a specific one instance to illustrate each step.

Step 1: Alice measures qubit 7 of the resource state and A_1 in the bell basis. If Alice measures $|\psi_+\rangle$, the remainder state is

$$\frac{1}{4\sqrt{2}}(|000\rangle|\psi_+\rangle\frac{(\alpha_0|0\rangle_6|0\rangle_{A_2}+\alpha_1|0\rangle_6|1\rangle_{A_2}+\alpha_2|1\rangle_6|0\rangle_{A_2}+\alpha_3|1\rangle_6|1\rangle_{A_2})}{\sqrt{2}}+$$

$$|001\rangle|\phi_-\rangle\frac{(\alpha_0|1\rangle_6|0\rangle_{A_2}+\alpha_1|1\rangle_6|1\rangle_{A_2}+\alpha_2|0\rangle_6|0\rangle_{A_2}+\alpha_3|0\rangle_6|1\rangle_{A_2})}{\sqrt{2}}+$$

$$|010\rangle|\psi_-\rangle\frac{(-\alpha_0|1\rangle_6|0\rangle_{A_2}-\alpha_1|1\rangle_6|1\rangle_{A_2}-\alpha_2|0\rangle_6|0\rangle_{A_2}-\alpha_3|0\rangle_6|1\rangle_{A_2})}{\sqrt{2}}+$$

$$|011\rangle|\phi_+\rangle\frac{(\alpha_0|0\rangle_6|0\rangle_{A_2}+\alpha_1|0\rangle_6|1\rangle_{A_2}-\alpha_2|1\rangle_6|0\rangle_{A_2}-\alpha_3|1\rangle_6|1\rangle_{A_2})}{\sqrt{2}}+$$

$$|100\rangle|\phi_+\rangle\frac{(\alpha_0|1\rangle_6|0\rangle_{A_2}+\alpha_1|1\rangle_6|1\rangle_{A_2}+\alpha_2|0\rangle_6|0\rangle_{A_2}+\alpha_3|0\rangle_6|1\rangle_{A_2})}{\sqrt{2}}+$$

$$|101\rangle|\psi_-\rangle\frac{(-\alpha_0|0\rangle_6|0\rangle_{A_2}-\alpha_1|0\rangle_6|1\rangle_{A_2}-\alpha_2|1\rangle_6|0\rangle_{A_2}-\alpha_3|1\rangle_6|1\rangle_{A_2})}{\sqrt{2}}+$$

$$|110\rangle|\phi_-\rangle\frac{(-\alpha_0|0\rangle_6|0\rangle_{A_2}-\alpha_1|0\rangle_6|1\rangle_{A_2}+\alpha_2|1\rangle_6|0\rangle_{A_2}+\alpha_3|1\rangle_6|1\rangle_{A_2})}{\sqrt{2}}+$$

$$|111\rangle|\psi_+\rangle\frac{(-\alpha_0|1\rangle_6|0\rangle_{A_2}-\alpha_1|1\rangle_6|1\rangle_{A_2}+\alpha_2|0\rangle_6|0\rangle_{A_2}+\alpha_3|0\rangle_6|1\rangle_{A_2})}{\sqrt{2}})|\phi\rangle_B$$

Alice communicates her result to Bob using a classical channel.

Step 2: Bob measures qubit 2 of the resource state and B_1 in the bell basis. If Bob Measures $|\psi_+\rangle$, the remainder state is

$$\frac{1}{2\sqrt{2}}(|000\rangle\frac{(\alpha_0|0\rangle_6|0\rangle_{A_2}+\alpha_1|0\rangle_6|1\rangle_{A_2}+\alpha_2|1\rangle_6|0\rangle_{A_2}+\alpha_3|1\rangle_6|1\rangle_{A_2})}{\sqrt{2}}$$

$$\times\frac{1}{\sqrt{2}}(\beta_0|00\rangle+\beta_1|01\rangle+\beta_2|10\rangle+\beta_3|11\rangle)_{4,B_2}+|001\rangle$$

$$\times\frac{(\alpha_0|1\rangle_6|0\rangle_{A_2}+\alpha_1|1\rangle_6|1\rangle_{A_2}+\alpha_2|0\rangle_6|0\rangle_{A_2}+\alpha_3|0\rangle_6|1\rangle_{A_2})}{\sqrt{2}}$$

$$\times\frac{1}{\sqrt{2}}(\beta_0|00\rangle+\beta_1|01\rangle+\beta_2|10\rangle+\beta_3|11\rangle)_{4,B_2}+|010\rangle$$

$$\times\frac{(\alpha_0|0\rangle_6|0\rangle_{A_2}+\alpha_1|0\rangle_6|1\rangle_{A_2}-\alpha_2|1\rangle_6|0\rangle_{A_2}-\alpha_3|1\rangle_6|1\rangle_{A_2})}{\sqrt{2}}$$

$$\times\frac{1}{\sqrt{2}}(\beta_0|00\rangle+\beta_1|01\rangle\ \beta_2|10\rangle\ \beta_3|11\rangle)_{4,B_2}+|011\rangle|$$

$$\times\frac{(\alpha_0|0\rangle_6|0\rangle_{A_2}+\alpha_1|0\rangle_6|1\rangle_{A_2}-\alpha_2|1\rangle_6|0\rangle_{A_2}-\alpha_3|1\rangle_6|1\rangle_{A_2})}{\sqrt{2}}$$

$$+\frac{1}{\sqrt{2}}(\beta_0|10\rangle+\beta_1|11\rangle\ \beta_2|00\rangle\ \beta_3|01\rangle)_{4,B_2}+|100\rangle$$

$$\times\frac{(\alpha_0|1\rangle_6|0\rangle_{A_2}+\alpha_1|1\rangle_6|1\rangle_{A_2}+\alpha_2|0\rangle_6|0\rangle_{A_2}+\alpha_3|0\rangle_6|1\rangle_{A_2})}{\sqrt{2}}$$

$$\times\frac{1}{\sqrt{2}}(\beta_0|10\rangle+\beta_1|11\rangle+\beta_2|00\rangle+\beta_3|01\rangle)_{4,B_2}+|101\rangle$$

$$\times\frac{(-\alpha_0|0\rangle_6|0\rangle_{A_2}-\alpha_1|0\rangle_6|1\rangle_{A_2}-\alpha_2|1\rangle_6|0\rangle_{A_2}-\alpha_3|1\rangle_6|1\rangle_{A_2})}{\sqrt{2}}$$

$$\times\frac{1}{\sqrt{2}}(\beta_0|00\rangle+\beta_1|01\rangle\ -\beta_2|10\rangle-\beta_3|11\rangle)_{4,B_2}+|110\rangle$$
$$\times\frac{(-\alpha_0|0\rangle_6|0\rangle_{A_2}-\alpha_1|0\rangle_6|1\rangle_{A_2}+\alpha_2|1\rangle_6|0\rangle_{A_2}+\alpha_3|1\rangle_6|1\rangle_{A_2})}{\sqrt{2}}$$
$$\times\frac{1}{\sqrt{2}}(\beta_0|10\rangle+\beta_1|11\rangle-\beta_2|10\rangle-\beta_3|01\rangle)_{4,B_2}+|111\rangle \quad .$$
$$\times\frac{(-\alpha_0|1\rangle_6|0\rangle_{A_2}-\alpha_1|1\rangle_6|1\rangle_{A_2}+\alpha_2|0\rangle_6|0\rangle_{A_2}+\alpha_3|0\rangle_6|1\rangle_{A_2})}{\sqrt{2}}$$
$$\times\frac{1}{\sqrt{2}}(\beta_0|00\rangle+\beta_1|01\rangle+\beta_2|10\rangle+\beta_3|11\rangle)_{4,B_2})$$

Bob communicates his result via a classical channel to Alice.

Step 3: Charlie, Alice and Bob measure their qubits in the Z-basis. Let us say they all measure 0, we have the

$$\frac{1}{2}(|000\rangle\frac{(\alpha_0|0\rangle_6|0\rangle_{A_2}+\alpha_1|0\rangle_6|1\rangle_{A_2}+\alpha_2|1\rangle_6|0\rangle_{A_2}+\alpha_3|1\rangle_6|1\rangle_{A_2})}{\sqrt{2}}$$
$$\times\frac{1}{\sqrt{2}}(\beta_0|00\rangle+\beta_1|01\rangle+\beta_2|10\rangle+\beta_3|11\rangle)_{4,B_2}) \tag{A9}$$

Step 4: Let Alice apply a CNOT with A_2 as control and qubit 1 as target, and let Bob apply a CNOT with and B_2 as control and qubit 3 as target, to get

$$\frac{1}{2\sqrt{2}}(|0\rangle\frac{(\alpha_0|000\rangle+\alpha_1|101\rangle_{A_2}+\alpha_2|010\rangle+\alpha_3|111\rangle)_{1,6,A_2}}{\sqrt{2}}$$
$$\times\frac{1}{\sqrt{2}}(\beta_0|000\rangle+\beta_1|101\rangle+\beta_2|010\rangle+\beta_3|111\rangle)_{3,4,B_2} \tag{A10}$$

Step 4: Alice and Bob measure their qubits A_2 and B_2 in the X-basis. Let us say they obtain the state $|+\rangle=\frac{1}{\sqrt{2}}(|0\rangle+|1\rangle)$, then the composite state is given by

$$\frac{1}{2}(|0\rangle\frac{(\alpha_0|00\rangle+\alpha_1|10\rangle_{A_2}+\alpha_2|01\rangle+\alpha_3|11\rangle)_{1,6}}{\sqrt{2}}$$
$$\times\frac{1}{\sqrt{2}}(\beta_0|00\rangle+\beta_1|10\rangle+\beta_2|01\rangle+\beta_3|11\rangle)_{3,4} \tag{A11}$$

Step 5: We apply unitary transformations to the composite state to now get Alice's initial arbitrary state in Bob's terminal and Bob's initial arbitrary state in Alice's terminal. In this instance, the unitary transformation is simply $I\otimes I\otimes I\otimes I$ with I being the identity matrix.

B.3 Quantum teleportation of arbitrary three-qubit state

The seven-qubit resource state can be used for the perfect linear teleportation of an arbitrary three qubit state. In this case, Alice possesses qubits 1, 2, 3, 4 and 5, and the 6th and 7th particles belong to Bob. Alice wants to transport an arbitrary state $|\psi^{(3)}\rangle=a|000\rangle+b|001\rangle+c|010\rangle+d|011\rangle+e|100\rangle+f|101\rangle+g|110\rangle+h|111\rangle$ to Bob. Using the decomposition given in *Supplementary Material*, the states possessed by, and the unitary transforms to be performed by, Bob have been recorded, to accomplish the teleportation of an arbitrary three-qubit state. A point to note here is that we get the GHZ state for $a = h = \frac{1}{\sqrt{2}},\ b = c = d = e = f = g = 0$ and the W state for $b = c = e = \frac{1}{\sqrt{3}},\ a = d = f = g = h = 0$.

The teleportation of an arbitrary three-qubit state using our resource-state has as the initial composite state,

$$
\begin{aligned}
|\Gamma_7^{(3)}\rangle &= |\psi^{(3)}\rangle \otimes |\Gamma_7\rangle \\
&= aA_{000}|000\rangle + aA_{001}|001\rangle + aA_{010}|010\rangle + aA_{011}|011\rangle \\
&+ aA_{100}|100\rangle + aA_{101}|101\rangle + aA_{110}|110\rangle + aA_{111}|111\rangle \\
&+ bB_{000}|000\rangle + bB_{001}|001\rangle + bB_{010}|010\rangle + bB_{011}|011\rangle \\
&+ bB_{100}|100\rangle + bB_{101}|101\rangle + bB_{110}|110\rangle + bB_{111}|111\rangle \\
&+ cC_{000}|000\rangle + cC_{001}|001\rangle + cC_{010}|010\rangle + cC_{011}|011\rangle \\
&+ cC_{100}|100\rangle + cC_{101}|101\rangle + cC_{110}|110\rangle + cC_{111}|111\rangle \\
&+ dD_{000}|000\rangle + dD_{001}|001\rangle + dD_{010}|010\rangle + dD_{011}|011\rangle \\
&+ dD_{100}|100\rangle + dD_{101}|101\rangle + dD_{110}|110\rangle + dD_{111}|111\rangle \qquad \text{(A12)} \\
&+ eE_{000}|000\rangle + eE_{001}|001\rangle + eE_{010}|010\rangle + eE_{011}|011\rangle \\
&+ eE_{100}|100\rangle + eE_{101}|101\rangle + eE_{110}|110\rangle + eE_{111}|111\rangle \\
&+ fF_{000}|000\rangle + fF_{001}|001\rangle + fF_{010}|010\rangle + fF_{011}|011\rangle \\
&+ fF_{100}|100\rangle + fF_{101}|101\rangle + fF_{110}|110\rangle + fB_{111}|111\rangle \\
&+ gG_{000}|000\rangle + gG_{001}|001\rangle + gG_{010}|010\rangle + gG_{011}|011\rangle \\
&+ gG_{100}|100\rangle + gG_{101}|101\rangle + gG_{110}|110\rangle + gG_{111}|111\rangle \\
&+ hH_{000}|000\rangle + hH_{001}|001\rangle + hH_{010}|010\rangle + hH_{011}|011\rangle \\
&+ hH_{100}|100\rangle + hH_{101}|101\rangle + hH_{110}|110\rangle + hH_{111}|111\rangle
\end{aligned}
$$

with

$$
\begin{aligned}
|A_{000}\rangle &= |0000000\rangle + |0000101\rangle - |0001011\rangle + |0001110\rangle \\
|A_{001}\rangle &= |0000010\rangle + |0001001\rangle + |0001100\rangle - |0000111\rangle \\
|A_{010}\rangle &= |0000111\rangle - |0000010\rangle + |0001001\rangle + |0001100\rangle \\
|A_{011}\rangle &= |0000000\rangle + |0000101\rangle + |0001011\rangle - |0001110\rangle \\
|A_{100}\rangle &= |0000011\rangle + |0000110\rangle - |0001000\rangle + |0001101\rangle \\
|A_{101}\rangle &= |0000001\rangle - |0000100\rangle + |0001010\rangle + |0001111\rangle \\
|A_{110}\rangle &= |0000001\rangle - |0000100\rangle - |0001010\rangle - |0001111\rangle \\
|A_{111}\rangle &= |0001101\rangle - |0000011\rangle - |0000110\rangle - |0001000\rangle
\end{aligned}
$$

$$|B_{000}\rangle = |0010000\rangle + |0010101\rangle - |0011011\rangle + |0011110\rangle$$
$$|B_{001}\rangle = |0010010\rangle + |0011001\rangle + |0011100\rangle - |0010111\rangle$$
$$|B_{010}\rangle = |0010111\rangle - |0010010\rangle + |0011001\rangle + |0011100\rangle$$
$$|B_{011}\rangle = |0010000\rangle + |0010101\rangle + |0011011\rangle - |0011110\rangle$$
$$|B_{100}\rangle = |0010011\rangle + |0010110\rangle - |0011000\rangle + |0011101\rangle$$
$$|B_{101}\rangle = |0010001\rangle - |0010100\rangle + |0011010\rangle + |0011111\rangle$$
$$|B_{110}\rangle = |0010001\rangle - |0010100\rangle - |0011010\rangle - |0011111\rangle$$
$$|B_{111}\rangle = |0011101\rangle - |0010011\rangle - |0010110\rangle - |0011000\rangle$$
$$|C_{000}\rangle = |0100000\rangle + |0100101\rangle - |0101011\rangle + |0101110\rangle$$
$$|C_{001}\rangle = |0100010\rangle + |0101001\rangle + |0101100\rangle - |0100111\rangle$$
$$|C_{010}\rangle = |0100111\rangle - |0100010\rangle + |0101001\rangle + |0101100\rangle$$
$$|C_{011}\rangle = |0100000\rangle + |0100101\rangle + |0101011\rangle - |0101110\rangle$$
$$|C_{100}\rangle = |0100011\rangle + |0100110\rangle - |0101000\rangle + |0101101\rangle$$
$$|C_{101}\rangle = |0100001\rangle - |0100100\rangle + |0101010\rangle + |0101111\rangle$$
$$|C_{110}\rangle = |0100001\rangle - |0100100\rangle - |0101010\rangle - |0101111\rangle$$
$$|C_{111}\rangle = |0101101\rangle - |0100011\rangle - |0100110\rangle - |0101000\rangle$$
$$|D_{000}\rangle = |0110000\rangle + |0110101\rangle - |0111011\rangle + |0111110\rangle$$
$$|D_{001}\rangle = |0110010\rangle + |0111001\rangle + |0111100\rangle - |0110111\rangle$$
$$|D_{010}\rangle = |0110111\rangle - |0110010\rangle + |0111001\rangle + |0111100\rangle$$
$$|D_{011}\rangle = |0110000\rangle + |0110101\rangle + |0111011\rangle - |0111110\rangle$$
$$|D_{100}\rangle = |0110011\rangle + |0110110\rangle - |0111000\rangle + |0111101\rangle$$
$$|D_{101}\rangle = |0110001\rangle - |0110100\rangle + |0111010\rangle + |0111111\rangle$$
$$|D_{110}\rangle = |0110001\rangle - |0110100\rangle - |0111010\rangle - |0111111\rangle$$
$$|D_{111}\rangle = |0111101\rangle - |0110011\rangle - |0110110\rangle - |0111000\rangle$$
$$|E_{000}\rangle = |1000000\rangle + |1000101\rangle - |1001011\rangle + |1001110\rangle$$
$$|E_{001}\rangle = |1000010\rangle + |1001001\rangle + |1001100\rangle - |1000111\rangle$$
$$|E_{010}\rangle = |1000111\rangle - |1000010\rangle + |0001001\rangle + |0001100\rangle$$
$$|E_{011}\rangle = |1000000\rangle + |1000101\rangle + |1001011\rangle - |1001110\rangle$$
$$|E_{100}\rangle = |1000011\rangle + |1000110\rangle - |1001000\rangle + |1001101\rangle$$
$$|E_{101}\rangle = |1000001\rangle - |1000100\rangle + |1001010\rangle + |1001111\rangle$$
$$|E_{110}\rangle = |1000001\rangle - |1000100\rangle - |1001010\rangle - |1001111\rangle$$
$$|E_{111}\rangle = |1001101\rangle - |1000011\rangle - |1000110\rangle - |1001000\rangle$$
$$|F_{000}\rangle = |1010000\rangle + |1010101\rangle - |1011011\rangle + |1011110\rangle$$
$$|F_{001}\rangle = |1010010\rangle + |1011001\rangle + |1011100\rangle - |1010111\rangle$$
$$|F_{010}\rangle = |1010111\rangle - |1010010\rangle + |1011001\rangle + |1011100\rangle$$
$$|F_{011}\rangle = |1010000\rangle + |1010101\rangle + |1011011\rangle - |1011110\rangle$$
$$|F_{100}\rangle = |1010011\rangle + |1010110\rangle - |1011000\rangle + |1011101\rangle$$
$$|F_{101}\rangle = |1010001\rangle - |1010100\rangle + |1011010\rangle + |1011111\rangle$$
$$|F_{110}\rangle = |1010001\rangle - |1010100\rangle - |1011010\rangle - |1011111\rangle$$
$$|F_{111}\rangle = |1011101\rangle - |1010011\rangle - |1010110\rangle - |1011000\rangle$$

$$
\begin{aligned}
|G_{000}\rangle &= |1100000\rangle + |1100101\rangle - |1101011\rangle + |1101110\rangle \\
|G_{001}\rangle &= |1100010\rangle + |1101001\rangle + |1101100\rangle - |1100111\rangle \\
|G_{010}\rangle &= |1100111\rangle - |1100010\rangle + |1101001\rangle + |1101100\rangle \\
|G_{011}\rangle &= |1100000\rangle + |1100101\rangle + |1101011\rangle - |1101110\rangle \\
|G_{100}\rangle &= |1100011\rangle + |1100110\rangle - |1101000\rangle + |1101101\rangle \\
|G_{101}\rangle &= |1100001\rangle - |1100100\rangle + |1101010\rangle + |1101111\rangle \\
|G_{110}\rangle &= |1100001\rangle - |1100100\rangle - |1101010\rangle - |1101111\rangle \\
|G_{111}\rangle &= |1101101\rangle - |1100011\rangle - |0000110\rangle - |0001000\rangle
\end{aligned}
$$

$$
\begin{aligned}
|H_{000}\rangle &= |1110000\rangle + |1110101\rangle - |1111011\rangle + |1111110\rangle \\
|H_{001}\rangle &= |1110010\rangle + |1111001\rangle + |1111100\rangle - |1110111\rangle \\
|H_{010}\rangle &= |1110111\rangle - |1110010\rangle + |1111001\rangle + |1111100\rangle \\
|H_{011}\rangle &= |1110000\rangle + |1110101\rangle + |1111011\rangle - |1111110\rangle \\
|H_{100}\rangle &= |1110011\rangle + |1110110\rangle - |1111000\rangle + |1111101\rangle \\
|H_{101}\rangle &= |1110001\rangle - |1110100\rangle + |1111010\rangle + |1111111\rangle \\
|H_{110}\rangle &= |1110001\rangle - |1110100\rangle - |1111010\rangle - |1111111\rangle \\
|H_{111}\rangle &= |1111101\rangle - |1110011\rangle - |1110110\rangle - |1111000\rangle
\end{aligned}
$$

An arbitrary three qubit state can be decomposed in terms of these basis-vectors,

$$
\begin{aligned}
&(a|000\rangle + b|001\rangle + c|010\rangle + d|011\rangle + e|100\rangle \\
&+f|101\rangle + g|110\rangle + h|111\rangle)|\Psi_7\rangle \\
&= \sum_{permutations} ((-1)^{I_1} A_{a_1a_2a_3} + (-1)^{I_2} B_{b_1b_2b_3} \\
&+(-1)^{I_3} C_{c_1c_2c_3} + (-1)^{I_4} D_{d_1d_2d_3} \\
&+(-1)^{I_5} E_{e_1e_2e_3} + (-1)^{I_6} F_{f_1f_2f_3} \\
&+(-1)^{I_7} G_{g_1g_2gs} + (-1)^{I_8} H_{h_1h_2h_3}) \\
&((-1)^{I_1} a|a_1a_2a_3\rangle + (-1)^{I_2} b|b_1b_2b_3\rangle \\
&+(-1)^{I_3} c|c_1c_2c_3\rangle + (-1)^{I_4} d|d_1d_2d_3\rangle \\
&+(-1)^{I_5} e|e_1e_2e_3\rangle + (-1)^{I_6} f|f_1f_2f_3\rangle \\
&+(-1)^{I_7} g|g_1g_2g_3\rangle + (-1)^{I_8} h|h_1h_2h_3\rangle)
\end{aligned}
\tag{A13}
$$

where $I_i(\mathrm{i} = 1, 2, 3, 4, 5, 6, 7, 8)$ can take values 0 or 1 independently, and $L_j(\mathrm{L} = \mathrm{a, b, c, d, e, f, g, h; j} = 1, 2, 3)$ can take values 0 or 1 independently. The summation is over all possible permutation states obtained.

The relevant transformations for the three-qubit teleportation are given in terms of the following basic operations:

Projection of ith component P_i:

$$
P_1 = \begin{pmatrix} 10 \\ 00 \end{pmatrix} \quad P_2 = \begin{pmatrix} 00 \\ 01 \end{pmatrix} \tag{A14}
$$

Flip and Projection of ith component F_i:

$$
F_1 = \begin{pmatrix} 01 \\ 00 \end{pmatrix} \quad F_2 = \begin{pmatrix} 00 \\ 10 \end{pmatrix} \tag{A15}
$$

State Obtained by Alice	Short-Hand Form of Transformation
$A_{000} + B_{001} + C_{010} + D_{011} + E_{100} + F_{101} + G_{110} + H_{111}$	$I_2 \otimes I_2 \otimes I_2$
$A_{000} - B_{001} + C_{010} + D_{011} + E_{100} + F_{101} + G_{110} + H_{111}$	$I_2 \otimes I_2 \otimes P_2 + \sigma_z \otimes P_1 \otimes P_1 + I_2 \otimes P_2 \otimes P_1$
$A_{000} + B_{001} + C_{010} - D_{011} + E_{100} + F_{101} + G_{110} + H_{111}$	$I_2 \otimes I_2 \otimes P_2 + I_2 \otimes P_1 \otimes P_1 + \sigma_z \otimes P_2 \otimes P_1$
$A_{000} + B_{001} + C_{010} + D_{011} + E_{100} - F_{101} + G_{110} + H_{111}$	$I_2 \otimes I_2 \otimes P_1 + \sigma_z \otimes P_1 \otimes P_2 + I_2 \otimes P_2 \otimes P_2$
$A_{000} + B_{001} + C_{010} + D_{011} + E_{100} + F_{101} + G_{110} - H_{111}$	$I_2 \otimes I_2 \otimes P_1 + \sigma_z \otimes P_2 \otimes P_2 + I_2 \otimes P_2 \otimes P_2$
$A_{000} - B_{001} + C_{010} - D_{011} + E_{100} + F_{101} + G_{110} + H_{111}$	$I_2 \otimes I_2 \otimes P_2 + \sigma_z \otimes P_1 \otimes P_1 + \sigma_z \otimes P_2 \otimes P_1$
$A_{000} - B_{001} + C_{010} + D_{011} + E_{100} - F_{101} + G_{110} + H_{111}$	$\sigma_z \otimes P_1 \otimes P_1 + I_2 \otimes P_2 \otimes P_1 + \sigma_z \otimes P_1 \otimes P_2 + I_2 \otimes P_2 \otimes P_2$
$A_{000} - B_{001} + C_{010} - D_{011} + E_{100} - F_{101} + G_{110} + H_{111}$	$\sigma_z \otimes P_1 \otimes P_1 + \sigma_z \otimes P_2 \otimes P_1 + \sigma_z \otimes P_1 \otimes P_2 + I_2 \otimes P_2 \otimes P_2$
$A_{000} + B_{001} + C_{010} - D_{011} + E_{100} - F_{101} + G_{110} + H_{111}$	$I_2 \otimes P_1 \otimes P_1 + \sigma_z \otimes P_2 \otimes P_1 + \sigma_z \otimes P_1 \otimes P_2 + I_2 \otimes P_2 \otimes P_2$
$A_{000} - B_{001} + C_{010} + D_{011} + E_{100} + F_{101} + G_{110} - H_{111}$	$\sigma_z \otimes P_1 \otimes P_1 + I_2 \otimes P_2 \otimes P_1 + P_2 \otimes P_1 \otimes P_2 + \sigma_z \otimes P_2 \otimes P_2$
$A_{000} + B_{001} + C_{010} - D_{011} + E_{100} + F_{101} + G_{110} - H_{111}$	$I_2 \otimes P_1 \otimes P_1 + \sigma_z \otimes P_2 \otimes P_1 + I_2 \otimes P_1 \otimes P_2 + \sigma_z \otimes P_2 \otimes P_2$
$A_{000} + B_{001} + C_{010} + D_{011} + E_{100} - F_{101} + G_{110} - H_{111}$	$I_2 \otimes P_1 \otimes P_1 + I_2 \otimes P_2 \otimes P_1 + \sigma_z \otimes P_1 \otimes P_2 + \sigma_z \otimes P_2 \otimes P_2$
$A_{000} - B_{001} + C_{010} + D_{011} + E_{100} - F_{101} + G_{110} - H_{111}$	$\sigma_z \otimes P_1 \otimes P_1 + I_2 \otimes P_2 \otimes P_1 + \sigma_z \otimes P_1 \otimes P_2 + \sigma_z \otimes P_2 \otimes P_2$
$A_{000} + B_{001} + C_{010} + D_{011} + E_{100} - F_{101} + G_{110} - H_{111}$	$I_2 \otimes P_1 \otimes P_1 + I_2 \otimes P_2 \otimes P_1 + \sigma_z \otimes P_1 \otimes P_2 + \sigma_z \otimes P_2 \otimes P_2$
$A_{000} - B_{001} + C_{010} - D_{011} + E_{100} + F_{101} + G_{110} - H_{111}$	$\sigma_z \otimes P_1 \otimes P_1 + \sigma_z \otimes P_2 \otimes P_1 + I_2 \otimes P_1 \otimes P_2 + \sigma_z \otimes P_2 \otimes P_2$
$A_{000} - B_{001} + C_{010} - D_{011} + E_{100} - F_{101} + G_{110} - H_{111}$	$\sigma_z \otimes P_1 \otimes P_1 + \sigma_z \otimes P_2 \otimes P_1 + \sigma_z \otimes P_1 \otimes P_2 + \sigma_z \otimes P_2 \otimes P_2$
$A_{000} + B_{001} - C_{010} + D_{011} + E_{100} + F_{101} + G_{110} + H_{111}$	$I_2 \otimes P_1 \otimes P_1 - \sigma_z \otimes P_2 \otimes P_1 + I_2 \otimes P_1 \otimes P_2 + I_2 \otimes P_2 \otimes P_2$
$A_{000} + B_{001} - C_{010} + D_{011} + E_{100} + F_{101} - G_{110} + H_{111}$	$I_2 \otimes P_1 \otimes P_1 - \sigma_z \otimes P_2 \otimes P_1 + I_2 \otimes P_1 \otimes P_2 - \sigma_z \otimes P_2 \otimes P_2$
$A_{000} + B_{001} - C_{010} - D_{011} + E_{100} + F_{101} + G_{110} + H_{111}$	$I_2 \otimes P_1 \otimes P_1 - I_2 \otimes P_2 \otimes P_1 + I_2 \otimes P_1 \otimes P_2 + I_2 \otimes P_2 \otimes P_2$
$A_{000} + B_{001} - C_{010} + D_{011} + E_{100} + F_{101} - G_{110} + H_{111}$	$I_2 \otimes P_1 \otimes P_1 - \sigma_z \otimes P_2 \otimes P_1 + I_2 \otimes P_1 \otimes P_2 - \sigma_z \otimes P_2 \otimes P_2$
$A_{000} + B_{001} - C_{010} + D_{011} + E_{100} + F_{101} + G_{110} - H_{111}$	$I_2 \otimes P_1 \otimes P_1 - \sigma_z \otimes P_2 \otimes P_1 + I_2 \otimes P_1 \otimes P_2 + \sigma_z \otimes P_2 \otimes P_2$
$A_{000} + B_{001} + C_{010} - D_{011} + E_{100} + F_{101} - G_{110} + H_{111}$	$I_2 \otimes P_1 \otimes P_1 + \sigma_z \otimes P_2 \otimes P_1 + I_2 \otimes P_1 \otimes P_2 + \sigma_z \otimes P_2 \otimes P_2$
$A_{000} + B_{001} + C_{010} + D_{011} + E_{100} + F_{101} - G_{110} - H_{111}$	$I_2 \otimes P_1 \otimes P_1 + I_2 \otimes P_2 \otimes P_1 + I_2 \otimes P_1 \otimes P_2 + \sigma_z \otimes P_2 \otimes P_2$
$A_{000} + B_{001} - C_{010} - D_{011} + E_{100} + F_{101} - G_{110} + H_{111}$	$I_2 \otimes P_1 \otimes P_1 - I_2 \otimes P_2 \otimes P_1 + I_2 \otimes P_1 \otimes P_2 - \sigma_z \otimes P_2 \otimes P_2$
$A_{000} + B_{001} - C_{010} - D_{011} + E_{100} + F_{101} + G_{110} - H_{111}$	$I_2 \otimes P_1 \otimes P_1 - I_2 \otimes P_2 \otimes P_1 + I_2 \otimes P_1 \otimes P_2 + \sigma_z \otimes P_2 \otimes P_2$
$A_{000} + B_{001} + C_{010} - D_{011} + E_{100} + F_{101} - G_{110} - H_{111}$	$I_2 \otimes P_1 \otimes P_1 + \sigma_z \otimes P_2 \otimes P_1 + I_2 \otimes P_1 \otimes P_2 - I_2 \otimes P_2 \otimes P_2$
$A_{000} + B_{001} - C_{010} + D_{011} + E_{100} + F_{101} - G_{110} - H_{111}$	$I_2 \otimes P_1 \otimes P_1 - \sigma_z \otimes P_2 \otimes P_1 + I_2 \otimes P_1 \otimes P_2 - I_2 \otimes P_2 \otimes P_2$
$A_{000} + B_{001} - C_{010} - D_{011} + E_{100} + F_{101} - G_{110} - H_{111}$	$I_2 \otimes P_1 \otimes P_1 - I_2 \otimes P_2 \otimes P_1 + I_2 \otimes P_1 \otimes P_2 - I_2 \otimes P_2 \otimes P_2$
$A_{000} + B_{001} + C_{010} + D_{011} - E_{100} + F_{101} + G_{110} + H_{111}$	$I_2 \otimes P_1 \otimes P_1 + I_2 \otimes P_2 \otimes P_1 - \sigma_z \otimes P_1 \otimes P_2 + I_2 \otimes P_2 \otimes P_2$
$A_{000} + B_{001} - C_{010} + D_{011} - E_{100} - F_{101} + G_{110} + H_{111}$	$\sigma_z \otimes P_1 \otimes P_1 - \sigma_z \otimes P_2 \otimes P_1 - I_2 \otimes P_1 \otimes P_2 + I_2 \otimes P_2 \otimes P_2$
$A_{000} + B_{001} + C_{010} + D_{011} - E_{100} - F_{101} + G_{110} + H_{111}$	$I_2 \otimes P_1 \otimes P_1 + I_2 \otimes P_2 \otimes P_1 - I_2 \otimes P_1 \otimes P_2 + I_2 \otimes P_2 \otimes P_2$
$A_{000} + B_{001} + C_{010} + D_{011} - E_{100} + F_{101} + G_{110} - H_{111}$	$I_2 \otimes P_1 \otimes P_1 + I_2 \otimes P_2 \otimes P_1 - \sigma_z \otimes P_1 \otimes P_2 + \sigma_z \otimes P_2 \otimes P_2$
$A_{000} + B_{001} + C_{010} + D_{011} + E_{100} - F_{101} - G_{110} + H_{111}$	$I_2 \otimes P_1 \otimes P_1 + I_2 \otimes P_2 \otimes P_1 + \sigma_z \otimes P_1 \otimes P_2 - \sigma_z \otimes P_2 \otimes P_2$
$A_{000} + B_{001} + C_{010} + D_{011} + E_{100} + F_{101} - G_{110} - H_{111}$	$I_2 \otimes P_1 \otimes P_1 + I_2 \otimes P_2 \otimes P_1 + I_2 \otimes P_1 \otimes P_2 - I_2 \otimes P_2 \otimes P_2$
$A_{000} + B_{001} + C_{010} + D_{011} + E_{100} - F_{101} - G_{110} + H_{111}$	$I_2 \otimes P_1 \otimes P_1 + I_2 \otimes P_2 \otimes P_1 + \sigma_z \otimes P_1 \otimes P_2 - \sigma_z \otimes P_2 \otimes P_2$
$A_{000} + B_{001} + C_{010} + D_{011} - E_{100} + F_{101} - G_{110} - H_{111}$	$I_2 \otimes P_1 \otimes P_1 + I_2 \otimes P_2 \otimes P_1 - \sigma_z \otimes P_1 \otimes P_2 - I_2 \otimes P_2 \otimes P_2$
$A_{000} + B_{001} + C_{010} + D_{011} + E_{100} - F_{101} - G_{110} - H_{111}$	$I_2 \otimes P_1 \otimes P_1 + I_2 \otimes P_2 \otimes P_1 + \sigma_z \otimes P_1 \otimes P_2 - I_2 \otimes P_2 \otimes P_2$
$A_{000} + B_{001} + C_{010} + D_{011} - E_{100} - F_{101} - G_{110} - H_{111}$	$I_2 \otimes P_1 \otimes P_1 + I_2 \otimes P_2 \otimes P_1 - I_2 \otimes P_1 \otimes P_2 - I_2 \otimes P_2 \otimes P_2$
$A_{001} + B_{000} + C_{011} + D_{010} + E_{101} + F_{100} + G_{111} + H_{110}$	$\sigma_x \otimes I_2 \otimes I_2$
$-A_{001} + B_{000} + C_{011} + D_{010} + E_{101} + F_{100} + G_{111} + H_{110}$	$\sigma_x \otimes P_1 \otimes P_2 + \sigma_x \otimes P_2 \otimes P_2 + I\sigma_y \otimes P_1 \otimes P_1$
$A_{001} + B_{000} - C_{011} + D_{010} + E_{101} + F_{100} + G_{111} + H_{110}$	$\sigma_x \otimes P_1 \otimes I_2 + I\sigma_y \otimes P_2 \otimes P_1 + \sigma_x \otimes I_2 \otimes P_2$
$A_{001} + B_{000} + C_{011} + D_{010} - E_{101} + F_{100} + G_{111} + H_{110}$	$\sigma_x \otimes I_2 \otimes P_1 + I\sigma_y \otimes P_1 \otimes P_2 + \sigma_x \otimes P_2 \otimes P_2$
$A_{001} + B_{000} + C_{011} + D_{010} + E_{101} + F_{100} - G_{111} + H_{110}$	$\sigma_x \otimes I_2 \otimes P_1 + \sigma_x \otimes P_1 \otimes P_2 + I\sigma_y \otimes P_2 \otimes P_2$
$-A_{001} + B_{000} - C_{011} + D_{010} + E_{101} + F_{100} + G_{111} + H_{110}$	$I\sigma_y \otimes I_2 \otimes P_1 + \sigma_x \otimes I_2 \otimes P_2$

State Obtained by Alice	Short-Hand Form of Transformation
$-A_{001}+B_{000}+C_{011}+D_{010}-E_{101}+F_{100}+G_{111}+H_{110}$	$I\sigma_y\otimes P_1\otimes P_1+\sigma_x\otimes P_2\otimes P_1+I\sigma_y\otimes P_1\otimes P_2+\sigma_x\otimes P_2\otimes P_2$
$-A_{001}+B_{000}+C_{011}+D_{010}+E_{101}+F_{100}-G_{111}+H_{110}$	$I\sigma_y\otimes P_1\otimes P_1+\sigma_x\otimes P_2\otimes P_1+\sigma_x\otimes P_1\otimes P_2+I\sigma_y\otimes P_2\otimes P_2$
$A_{001}+B_{000}-C_{011}+D_{010}-E_{101}+F_{100}+G_{111}+H_{110}$	$\sigma_x\otimes P_1\otimes P_1+I\sigma_y\otimes P_2\otimes P_1+I\sigma_y\otimes P_1\otimes P_2+\sigma_x\otimes P_2\otimes P_2$
$A_{001}+B_{000}-C_{011}+D_{010}+E_{101}+F_{100}-G_{111}+H_{110}$	$\sigma_x\otimes P_1\otimes P_1+I\sigma_y\otimes P_2\otimes P_1+\sigma_x\otimes P_1\otimes P_2+I\sigma_y\otimes P_2\otimes P_2$
$A_{001}+B_{000}+C_{011}+D_{010}-E_{101}+F_{100}-G_{111}+H_{110}$	$\sigma_x\otimes P_1\otimes P_1+\sigma_x\otimes P_2\otimes P_1+I\sigma_y\otimes P_1\otimes P_2+I\sigma_y\otimes P_2\otimes P_2$
$-A_{001}+B_{000}-C_{011}+D_{010}-E_{101}+F_{100}+G_{111}+H_{110}$	$\sigma_x\otimes P_1\otimes P_1+I\sigma_y\otimes P_2\otimes P_1+I\sigma_y\otimes P_1\otimes P_2+\sigma_x\otimes P_2\otimes P_2$
$-A_{001}+B_{000}+C_{011}+D_{010}-E_{101}+F_{100}-G_{111}+H_{110}$	$I\sigma_y\otimes P_1\otimes P_1+\sigma_x\otimes P_2\otimes P_1+I\sigma_y\otimes P_1\otimes P_2+I\sigma_y\otimes P_2\otimes P_2$
$-A_{001}+B_{000}-C_{011}+D_{010}+E_{101}+F_{100}-G_{111}+H_{110}$	$I\sigma_y\otimes P_1\otimes P_1+I\sigma_y\otimes P_2\otimes P_1+\sigma_x\otimes P_1\otimes P_2+I\sigma_y\otimes P_2\otimes P_2$
$A_{001}+B_{000}-C_{011}+D_{010}-E_{101}+F_{100}-G_{111}+H_{110}$	$\sigma_x\otimes P_1\otimes P_1+I\sigma_y\otimes P_2\otimes P_1+I\sigma_y\otimes P_1\otimes P_2+I\sigma_y\otimes P_2\otimes P_2$
$-A_{001}+B_{000}-C_{011}+D_{010}-E_{101}+F_{100}-G_{111}+H_{110}$	$I\sigma_y\otimes P_1\otimes P_1+I\sigma_y\otimes P_2\otimes P_1+\sigma_x\otimes P_1\otimes P_2+I\sigma_y\otimes P_2\otimes P_2$
$A_{001}+B_{000}+C_{011}-D_{010}+E_{101}+F_{100}+G_{111}+H_{110}$	$\sigma_x\otimes P_1\otimes P_1-I\sigma_y\otimes P_2\otimes P_1+\sigma_x\otimes P_1\otimes P_2+\sigma_x\otimes P_2\otimes P_2$
$A_{001}+B_{000}+C_{011}+D_{010}+E_{101}+F_{100}+G_{111}-H_{110}$	$\sigma_x\otimes P_1\otimes P_1+\sigma_x\otimes P_2\otimes P_1+\sigma_x\otimes P_1\otimes P_2-I\sigma_y\otimes P_2\otimes P_2$
$A_{001}+B_{000}-C_{011}+D_{010}+E_{101}+F_{100}+G_{111}-H_{110}$	$\sigma_x\otimes P_1\otimes P_1+I\sigma_y\otimes P_2\otimes P_1+\sigma_x\otimes P_1\otimes P_2+I\sigma_y\otimes P_2\otimes P_2$
$A_{001}+B_{000}-C_{011}-D_{010}+E_{101}+F_{100}+G_{111}+H_{110}$	$\sigma_x\otimes P_1\otimes P_1-\sigma_x\otimes P_2\otimes P_1+\sigma_x\otimes P_1\otimes P_2+\sigma_x\otimes P_2\otimes P_2$
$A_{001}+B_{000}+C_{011}-D_{010}+E_{101}+F_{100}-G_{111}+H_{110}$	$\sigma_x\otimes P_1\otimes P_1-I\sigma_y\otimes P_2\otimes P_1+\sigma_x\otimes P_1\otimes P_2+I\sigma_y\otimes P_2\otimes P_2$
$A_{001}+B_{000}+C_{011}-D_{010}+E_{101}+F_{100}+G_{111}-H_{110}$	$\sigma_x\otimes P_1\otimes P_1-I\sigma_y\otimes P_2\otimes P_1+\sigma_x\otimes P_1\otimes P_2-I\sigma_y\otimes P_2\otimes P_2$
$A_{001}+B_{000}+C_{011}+D_{010}+E_{101}+F_{100}-G_{111}-H_{110}$	$\sigma_x\otimes P_1\otimes P_1+\sigma_x\otimes P_2\otimes P_1+I\sigma_y\otimes P_1\otimes P_2-\sigma_x\otimes P_2\otimes P_2$
$A_{001}+B_{000}-C_{011}-D_{010}+E_{101}+F_{100}-G_{111}+H_{110}$	$\sigma_x\otimes P_1\otimes P_1-\sigma_x\otimes P_2\otimes P_1+\sigma_x\otimes P_1\otimes P_2+I\sigma_y\otimes P_2\otimes P_2$
$A_{001}+B_{000}-C_{011}-D_{010}+E_{101}+F_{100}+G_{111}-H_{110}$	$\sigma_x\otimes P_1\otimes P_1-\sigma_x\otimes P_2\otimes P_1+\sigma_x\otimes P_1\otimes P_2-I\sigma_y\otimes P_2\otimes P_2$
$A_{001}+B_{000}-C_{011}+D_{010}+E_{101}+F_{100}-G_{111}-H_{110}$	$\sigma_x\otimes P_1\otimes P_1-\sigma_x\otimes P_2\otimes P_1+\sigma_x\otimes P_1\otimes P_2-I\sigma_y\otimes P_2\otimes P_2$
$A_{001}+B_{000}+C_{011}-D_{010}+E_{101}+F_{100}-G_{111}-H_{110}$	$\sigma_x\otimes P_1\otimes P_1-I\sigma_y\otimes P_2\otimes P_1+\sigma_x\otimes P_1\otimes P_2-\sigma_x\otimes P_2\otimes P_2$
$A_{001}+B_{000}-C_{011}-D_{010}+E_{101}+F_{100}-G_{111}-H_{110}$	$\sigma_x\otimes P_1\otimes P_1-\sigma_x\otimes P_2\otimes P_1+\sigma_x\otimes P_1\otimes P_2-\sigma_x\otimes P_2\otimes P_2$
$A_{001}+B_{000}+C_{011}+D_{010}+E_{101}-F_{100}+G_{111}+H_{110}$	$\sigma_x\otimes P_1\otimes P_1+\sigma_x\otimes P_2\otimes P_1-I\sigma_y\otimes P_1\otimes P_2+\sigma_x\otimes P_2\otimes P_2$
$A_{001}+B_{000}+C_{011}+D_{010}-E_{101}-F_{100}+G_{111}+H_{110}$	$\sigma_x\otimes P_1\otimes P_1+\sigma_x\otimes P_2\otimes P_1+I\sigma_y\otimes P_1\otimes P_2+\sigma_x\otimes P_2\otimes P_2$
$A_{001}+B_{000}+C_{011}+D_{010}-E_{101}+F_{100}-G_{111}+H_{110}$	$\sigma_x\otimes P_1\otimes P_1+\sigma_x\otimes P_2\otimes P_1+I\sigma_y\otimes P_1\otimes P_2+I\sigma_y\otimes P_2\otimes P_2$
$A_{001}+B_{000}+C_{011}+D_{010}-E_{101}+F_{100}+G_{111}-H_{110}$	$\sigma_x\otimes P_1\otimes P_1+\sigma_x\otimes P_2\otimes P_1+I\sigma_y\otimes P_1\otimes P_2-I\sigma_y\otimes P_2\otimes P_2$
$A_{001}+B_{000}+C_{011}+D_{010}+E_{101}-F_{100}-G_{111}+H_{110}$	$\sigma_x\otimes P_1\otimes P_1+\sigma_x\otimes P_2\otimes P_1-I\sigma_y\otimes P_1\otimes P_2+I\sigma_y\otimes P_2\otimes P_2$
$A_{001}+B_{000}+C_{011}+D_{010}+E_{101}-F_{100}+G_{111}-H_{110}$	$\sigma_x\otimes P_1\otimes P_1+\sigma_x\otimes P_2\otimes P_1-I\sigma_y\otimes P_1\otimes P_2-I\sigma_y\otimes P_2\otimes P_2$
$A_{001}+B_{000}+C_{011}+D_{010}-E_{101}-F_{100}-G_{111}+H_{110}$	$\sigma_x\otimes P_1\otimes P_1+\sigma_x\otimes P_2\otimes P_1-\sigma_x\otimes P_1\otimes P_2+I\sigma_y\otimes P_2\otimes P_2$
$A_{001}+B_{000}+C_{011}+D_{010}-E_{101}+F_{100}-G_{111}-H_{110}$	$\sigma_x\otimes P_1\otimes P_1+\sigma_x\otimes P_2\otimes P_1+I\sigma_y\otimes P_1\otimes P_2-\sigma_x\otimes P_2\otimes P_2$
$A_{001}+B_{000}+C_{011}+D_{010}+E_{101}-F_{100}-G_{111}-H_{110}$	$\sigma_x\otimes P_1\otimes P_1+\sigma_x\otimes P_2\otimes P_1-I\sigma_y\otimes P_1\otimes P_2-\sigma_x\otimes P_2\otimes P_2$
$+A_{001}+B_{000}+C_{011}+D_{010}-E_{101}-F_{100}-G_{111}-H_{110}$	$\sigma_x\otimes P_1\otimes P_1+\sigma_x\otimes P_2\otimes P_1-\sigma_x\otimes P_1\otimes P_2-\sigma_x\otimes P_2\otimes P_2$
$A_{010}-B_{011}+C_{000}+D_{001}+E_{110}+F_{111}+G_{100}+H_{101}$	$\sigma_z\otimes F_1\otimes P_1+I_2\otimes F_2\otimes P_1+I_2\otimes F_1\otimes P_2+I_2\otimes P_2\otimes P_2$
$A_{010}+B_{011}+C_{000}-D_{001}+E_{110}+F_{111}+G_{100}+H_{101}$	$I_2\otimes F_1\otimes P_1+\sigma_z\otimes F_2\otimes P_1+I_2\otimes F_1\otimes P_2+I_2\otimes P_2\otimes P_2$
$A_{010}+B_{011}+C_{000}+D_{001}+E_{110}-F_{111}+G_{100}+H_{101}$	$I_2\otimes F_1\otimes P_1+I_2\otimes F_2\otimes P_1+\sigma_z\otimes F_1\otimes P_2+I_2\otimes P_2\otimes P_2$
$A_{010}+B_{011}+C_{000}+D_{001}+E_{110}+F_{111}+G_{100}-H_{101}$	$\sigma_z\otimes F_1\otimes P_1+I_2\otimes F_2\otimes P_1+I_2\otimes F_1\otimes P_2+\sigma_z\otimes P_2\otimes P_2$
$A_{010}-B_{011}+C_{000}-D_{001}+E_{110}+F_{111}+G_{100}+H_{101}$	$\sigma_z\otimes F_1\otimes P_1+\sigma_z\otimes F_2\otimes P_1+I_2\otimes F_1\otimes P_2+I_2\otimes P_2\otimes P_2$
$A_{010}-B_{011}+C_{000}+D_{001}+E_{110}-F_{111}+G_{100}+H_{101}$	$\sigma_z\otimes F_1\otimes P_1+I_2\otimes F_2\otimes P_1+\sigma_z\otimes F_1\otimes P_2+I_2\otimes P_2\otimes P_2$
$A_{010}-B_{011}+C_{000}+D_{001}+E_{110}+F_{111}+G_{100}-H_{101}$	$\sigma_z\otimes F_1\otimes P_1+I_2\otimes F_2\otimes P_1+I_2\otimes F_1\otimes P_2+\sigma_z\otimes P_2\otimes P_2$
$A_{010}+B_{011}+C_{000}+D_{001}+E_{110}+F_{111}+G_{100}+H_{101}$	$I_2\otimes F_1\otimes P_1+I_2\otimes F_2\otimes P_1+I_2\otimes F_1\otimes P_2+I_2\otimes P_2\otimes P_2$
$A_{010}+B_{011}+C_{000}-D_{001}+E_{110}-F_{111}+G_{100}+H_{101}$	$I_2\otimes F_1\otimes P_1+\sigma_z\otimes F_2\otimes P_1+\sigma_z\otimes F_1\otimes P_2+I_2\otimes P_2\otimes P_2$
$A_{010}+B_{011}+C_{000}-D_{001}+E_{110}+F_{111}+G_{100}-H_{101}$	$I_2\otimes F_1\otimes P_1+\sigma_z\otimes F_2\otimes P_1+I_2\otimes F_1\otimes P_2+\sigma_z\otimes P_2\otimes P_2$
$A_{010}+B_{011}+C_{000}+D_{001}+E_{110}-F_{111}+G_{100}-H_{101}$	$I_2\otimes F_1\otimes P_1+I_2\otimes F_2\otimes P_1+\sigma_z\otimes F_1\otimes P_2+\sigma_z\otimes P_2\otimes P_2$

State Obtained by Alice	Short-Hand Form of Transformation
$A_{010} - B_{011} + C_{000} - D_{001} + E_{110} - F_{111} + G_{100} + H_{101}$	$\sigma_z \otimes F_1 \otimes P_1 + \sigma_z \otimes F_2 \otimes P_1 + I_2 \otimes F_1 \otimes P_2 + I_2 \otimes P_2 \otimes P_2$
$A_{010} - B_{011} + C_{000} - D_{001} + E_{110} + F_{111} + G_{100} - H_{101}$	$\sigma_z \otimes F_1 \otimes P_1 + \sigma_z \otimes F_2 \otimes P_1 + I_2 \otimes F_1 \otimes P_2 + \sigma_z \otimes P_2 \otimes P_2$
$A_{010} + B_{011} + C_{000} - D_{001} + E_{110} - F_{111} + G_{100} - H_{101}$	$I_2 \otimes F_1 \otimes P_1 + \sigma_z \otimes F_2 \otimes P_1 + \sigma_z \otimes F_1 \otimes P_2 + \sigma_z \otimes P_2 \otimes P_2$
$A_{010} - B_{011} + C_{000} + D_{001} + E_{110} - F_{111} + G_{100} - H_{101}$	$\sigma_z \otimes F_1 \otimes P_1 + I_2 \otimes F_2 \otimes P_1 + \sigma_z \otimes F_1 \otimes P_2 + \sigma_z \otimes P_2 \otimes P_2$
$A_{010} - B_{011} + C_{000} - D_{001} + E_{110} - F_{111} + G_{100} - H_{101}$	$\sigma_z \otimes F_1 \otimes P_1 + \sigma_z \otimes F_2 \otimes P_1 + \sigma_z \otimes P_2 \otimes P_2 + \sigma_z \otimes P_2 \otimes P_2$
$-A_{010} + B_{011} + C_{000} + D_{001} + E_{110} + F_{111} + G_{100} + H_{101}$	$\sigma_z \otimes F_1 \otimes P_1 + I_2 \otimes F_2 \otimes P_1 + I_2 \otimes F_1 \otimes P_2 + I_2 \otimes P_2 \otimes P_2$
$A_{010} + B_{011} + C_{000} + D_{001} - E_{110} + F_{111} + G_{100} + H_{101}$	$I_2 \otimes F_1 \otimes P_1 + I_2 \otimes F_2 \otimes P_1 - \sigma_z \otimes F_1 \otimes P_2 + I_2 \otimes P_2 \otimes P_2$
$-A_{010} - B_{011} + C_{000} + D_{001} + E_{110} + F_{111} + G_{100} + H_{101}$	$-I_2 \otimes F_1 \otimes P_1 + I_2 \otimes F_2 \otimes P_1 + I_2 \otimes F_1 \otimes P_2 + I_2 \otimes P_2 \otimes P_2$
$-A_{010} + B_{011} + C_{000} + D_{001} - E_{110} + F_{111} + G_{100} + H_{101}$	$-\sigma_z \otimes F_1 \otimes P_1 + I_2 \otimes F_2 \otimes P_1 - \sigma_z \otimes F_1 \otimes P_2 + I_2 \otimes P_2 \otimes P_2$
$-A_{010} + B_{011} + C_{000} + D_{001} + E_{110} - F_{111} + G_{100} + H_{101}$	$-\sigma_z \otimes F_1 \otimes P_1 + I_2 \otimes F_2 \otimes P_1 + \sigma_z \otimes F_1 \otimes P_2 + I_2 \otimes P_2 \otimes P_2$
$A_{010} - B_{011} + C_{000} + D_{001} - E_{110} + F_{111} + G_{100} + H_{101}$	$-\sigma_z \otimes F_1 \otimes P_1 + I_2 \otimes F_2 \otimes P_1 - \sigma_z \otimes F_1 \otimes P_2 + I_2 \otimes P_2 \otimes P_2$
$A_{010} + B_{011} + C_{000} + D_{001} - E_{110} - F_{111} + G_{100} + H_{101}$	$I_2 \otimes F_1 \otimes P_1 + I_2 \otimes F_2 \otimes P_1 - I_2 \otimes F_1 \otimes P_2 + I_2 \otimes P_2 \otimes P_2$
$-A_{010} - B_{011} + C_{000} + D_{001} - E_{110} + F_{111} + G_{100} + H_{101}$	$-I_2 \otimes F_1 \otimes P_1 + I_2 \otimes F_2 \otimes P_1 - \sigma_z \otimes F_1 \otimes P_2 + I_2 \otimes P_2 \otimes P_2$
$-A_{010} + B_{011} + C_{000} + D_{001} - E_{110} - F_{111} + G_{100} + H_{101}$	$-\sigma_z \otimes F_1 \otimes P_1 + I_2 \otimes F_2 \otimes P_1 - I_2 \otimes F_1 \otimes P_2 + I_2 \otimes P_2 \otimes P_2$
$A_{010} - B_{011} + C_{000} + D_{001} - E_{110} - F_{111} + G_{100} + H_{101}$	$\sigma_z \otimes F_1 \otimes P_1 + I_2 \otimes F_2 \otimes P_1 - I_2 \otimes F_1 \otimes P_2 + I_2 \otimes P_2 \otimes P_2$
$-A_{010} - B_{011} + C_{000} + D_{001} + E_{110} - F_{111} + G_{100} + H_{101}$	$-I_2 \otimes F_1 \otimes P_1 + I_2 \otimes F_2 \otimes P_1 + \sigma_z \otimes F_1 \otimes P_2 + I_2 \otimes P_2 \otimes P_2$
$-A_{010} - B_{011} + C_{000} + D_{001} - E_{110} - F_{111} + G_{100} + H_{101}$	$-I_2 \otimes F_1 \otimes P_1 + I_2 \otimes F_2 \otimes P_1 - I_2 \otimes F_1 \otimes P_2 + I_2 \otimes P_2 \otimes P_2$
$A_{010} + B_{011} + C_{000} + D_{001} + E_{110} + F_{111} - G_{100} + H_{101}$	$I_2 \otimes F_1 \otimes P_1 + I_2 \otimes F_2 \otimes P_1 + I_2 \otimes F_1 \otimes P_2 - \sigma_z \otimes P_2 \otimes P_2$
$A_{010} + B_{011} + C_{000} + D_{001} - E_{110} + F_{111} - G_{100} + H_{101}$	$I_2 \otimes F_1 \otimes P_1 + I_2 \otimes F_2 \otimes P_1 - \sigma_z \otimes F_1 \otimes P_2 - \sigma_z \otimes P_2 \otimes P_2$
$A_{010} + B_{011} + C_{000} + D_{001} - E_{110} + F_{111} + G_{100} - H_{101}$	$I_2 \otimes F_1 \otimes P_1 + I_2 \otimes F_2 \otimes P_1 - \sigma_z \otimes F_1 \otimes P_2 + \sigma_z \otimes P_2 \otimes P_2$
$A_{010} + B_{011} + C_{000} + D_{001} + E_{110} - F_{111} - G_{100} + H_{101}$	$I_2 \otimes F_1 \otimes P_1 + I_2 \otimes F_2 \otimes P_1 + \sigma_z \otimes F_1 \otimes P_2 - \sigma_z \otimes P_2 \otimes P_2$
$A_{010} + B_{011} + C_{000} + D_{001} + E_{110} - F_{111} + G_{100} - H_{101}$	$I_2 \otimes F_1 \otimes P_1 + I_2 \otimes F_2 \otimes P_1 + \sigma_z \otimes F_1 \otimes P_2 + \sigma_z \otimes P_2 \otimes P_2$
$A_{010} + B_{011} + C_{000} + D_{001} - E_{110} - F_{111} - G_{100} + H_{101}$	$I_2 \otimes F_1 \otimes P_1 + I_2 \otimes F_2 \otimes P_1 - I_2 \otimes F_1 \otimes P_2 - \sigma_z \otimes P_2 \otimes P_2$
$A_{010} + B_{011} + C_{000} + D_{001} - E_{110} + F_{111} - G_{100} - H_{101}$	$I_2 \otimes F_1 \otimes P_1 + I_2 \otimes F_2 \otimes P_1 - \sigma_z \otimes F_1 \otimes P_2 - I_2 \otimes P_2 \otimes P_2$
$A_{010} + B_{011} + C_{000} + D_{001} + E_{110} - F_{111} - G_{100} - H_{101}$	$I_2 \otimes F_1 \otimes P_1 + I_2 \otimes F_2 \otimes P_1 + \sigma_z \otimes F_1 \otimes P_2 - I_2 \otimes P_2 \otimes P_2$
$A_{010} + B_{011} + C_{000} + D_{001} - E_{110} - F_{111} + G_{100} - H_{101}$	$I_2 \otimes F_1 \otimes P_1 + I_2 \otimes F_2 \otimes P_1 - I_2 \otimes F_1 \otimes P_2 + \sigma_z \otimes P_2 \otimes P_2$
$A_{010} + B_{011} + C_{000} + D_{001} - E_{110} - F_{111} - G_{100} - H_{101}$	$I_2 \otimes F_1 \otimes P_1 + I_2 \otimes F_2 \otimes P_1 - I_2 \otimes F_1 \otimes P_2 - I_2 \otimes P_2 \otimes P_2$
$A_{011} + B_{010} + C_{001} + D_{000} - E_{111} - F_{110} - G_{101} - H_{100}$	$\sigma_x \otimes \sigma_x \otimes P_1 - \sigma_x \otimes \sigma_x \otimes P_2$
$A_{011} + B_{010} + C_{001} + D_{000} + E_{111} + F_{110} + G_{101} + H_{100}$	$\sigma_x \otimes \sigma_x \otimes I_2$
$-A_{011} + B_{010} + C_{001} + D_{000} + E_{111} + F_{110} + G_{101} + H_{100}$	$I\sigma_y \otimes F_1 \otimes P_1 + \sigma_x \otimes F_2 \otimes P_1 + \sigma_x \otimes \sigma_x \otimes P_2$
$A_{011} + B_{010} - C_{001} + D_{000} + E_{111} + F_{110} + G_{101} + H_{100}$	$\sigma_x \otimes F_1 \otimes P_1 + I\sigma_y \otimes F_2 \otimes P_1 - I\sigma_y \otimes F_1 \otimes P_2 + \sigma_x \otimes F_2 \otimes P_2$
$A_{011} + B_{010} + C_{001} + D_{000} - E_{111} + F_{110} + G_{101} + H_{100}$	$\sigma_x \otimes \sigma_x \otimes P_1 + I\sigma_y \otimes F_1 \otimes P_2 + \sigma_x \otimes F_2 \otimes P_2$
$A_{011} + B_{010} + C_{001} + D_{000} + E_{111} + F_{110} - G_{101} + H_{100}$	$\sigma_x \otimes \sigma_x \otimes P_1 + \sigma_x \otimes F_1 \otimes P_2 + I\sigma_y \otimes F_2 \otimes P_2$
$-A_{011} + B_{010} - C_{001} + D_{000} + E_{111} + F_{110} + G_{101} + H_{100}$	$I\sigma_y \otimes \sigma_x \otimes P_1 + \sigma_x \otimes F_1 \otimes P_2 + \sigma_x \otimes F_2 \otimes P_2$
$-A_{011} + B_{010} + C_{001} + D_{000} - E_{111} + F_{110} + G_{101} + H_{100}$	$I\sigma_y \otimes F_1 \otimes P_1 + \sigma_x \otimes F_2 \otimes P_1 + I\sigma_y \otimes F_1 \otimes P_2 + \sigma_x \otimes F_2 \otimes P_2$
$-A_{011} + B_{010} + C_{001} + D_{000} + E_{111} + F_{110} - G_{101} + H_{100}$	$I\sigma_y \otimes F_1 \otimes P_1 + \sigma_x \otimes F_2 \otimes P_1 + \sigma_x \otimes F_1 \otimes P_2 + I\sigma_y \otimes F_2 \otimes P_2$
$A_{011} + B_{010} - C_{001} + D_{000} - E_{111} + F_{110} + G_{101} + H_{100}$	$\sigma_x \otimes F_1 \otimes P_1 + I\sigma_y \otimes F_2 \otimes P_1 + I\sigma_y \otimes F_1 \otimes P_2 + \sigma_x \otimes F_2 \otimes P_2$
$A_{011} + B_{010} - C_{001} + D_{000} + E_{111} + F_{110} - G_{101} + H_{100}$	$\sigma_x \otimes F_1 \otimes P_1 + I\sigma_y \otimes F_2 \otimes P_1 + \sigma_x \otimes F_1 \otimes P_2 + I\sigma_y \otimes F_2 \otimes P_2$
$A_{011} + B_{010} + C_{001} + D_{000} - E_{111} + F_{110} - G_{101} + H_{100}$	$\sigma_x \otimes \sigma_x \otimes P_1 + I\sigma_y \otimes F_1 \otimes P_2 + I\sigma_y \otimes F_2 \otimes P_2$
$-A_{011} + B_{010} - C_{001} + D_{000} - E_{111} + F_{110} + G_{101} + H_{100}$	$I\sigma_y \otimes \sigma_x \otimes P_1 + I\sigma_y \otimes F_1 \otimes P_2 + \sigma_x \otimes F_2 \otimes P_2$
$-A_{011} + B_{010} - C_{001} + D_{000} + E_{111} + F_{110} - G_{101} + H_{100}$	$I\sigma_y \otimes \sigma_x \otimes P_1 + \sigma_x \otimes F_1 \otimes P_2 + I\sigma_y \otimes F_2 \otimes P_2$
$-A_{011} + B_{010} + C_{001} + D_{000} - E_{111} + F_{110} - G_{101} + H_{100}$	$I\sigma_y \otimes F_1 \otimes P_1 + \sigma_x \otimes F_2 \otimes P_1 + I\sigma_y \otimes \sigma_x \otimes P_2$
$A_{011} + B_{010} - C_{001} + D_{000} - E_{111} + F_{110} - G_{101} + H_{100}$	$\sigma_x \otimes F_1 \otimes P_1 + I\sigma_y \otimes F_2 \otimes P_1 + I\sigma_y \otimes F_1 \otimes P_2 + I\sigma_y \otimes F_2 \otimes P_2$

State Obtained by Alice	Short-Hand Form of Transformation
$-A_{011}+B_{010}-C_{001}+D_{000}-E_{111}+F_{110}-G_{101}+H_{100}$	$I\sigma_y\otimes\sigma_x\otimes P_1+I\sigma_y\otimes F_1\otimes P_2+I\sigma_y\otimes F_2\otimes P_2$
$A_{011}-B_{010}+C_{001}+D_{000}+E_{111}+F_{110}+G_{101}+H_{100}$	$-I\sigma_y\otimes F_1\otimes P_1+\sigma_x\otimes F_2\otimes P_1+\sigma_x\otimes\sigma_x\otimes P_2$
$A_{011}+B_{010}+C_{001}+D_{000}+E_{111}-F_{110}+G_{101}+H_{100}$	$\sigma_x\otimes\sigma_x\otimes P_1-I\sigma_y\otimes F_1\otimes P_2+\sigma_x\otimes F_2\otimes P_2$
$-A_{011}-B_{010}+C_{001}+D_{000}+E_{111}+F_{110}+G_{101}+H_{100}$	$\sigma_x\otimes I\sigma_y\otimes P_1+\sigma_x\otimes F_1\otimes P_2+\sigma_x\otimes F_2\otimes P_2$
$-A_{011}+B_{010}+C_{001}+D_{000}+E_{111}-F_{110}+G_{101}+H_{100}$	$I\sigma_y\otimes F_1\otimes P_1+\sigma_x\otimes F_2\otimes P_1-I\sigma_y\otimes F_1\otimes P_2+\sigma_x\otimes F_2\otimes P_2$
$A_{011}-B_{010}+C_{001}+D_{000}-E_{111}+F_{110}+G_{101}+H_{100}$	$-I\sigma_y\otimes F_1\otimes P_1+\sigma_x\otimes F_2\otimes P_1+I\sigma_y\otimes F_1\otimes P_2+\sigma_x\otimes F_2\otimes P_2$
$A_{011}-B_{010}+C_{001}+D_{000}+E_{111}-F_{110}+G_{101}+H_{100}$	$-I\sigma_y\otimes F_1\otimes P_1+\sigma_x\otimes F_2\otimes P_1-I\sigma_y\otimes F_1\otimes P_2+\sigma_x\otimes F_2\otimes P_2$
$A_{011}+B_{010}+C_{001}+D_{000}-E_{111}-F_{110}+G_{101}+H_{100}$	$\sigma_x\otimes\sigma_x\otimes P_1+\sigma_x\otimes I\sigma_y\otimes P_2$
$-A_{011}-B_{010}+C_{001}+D_{000}-E_{111}+F_{110}+G_{101}+H_{100}$	$\sigma_x\otimes I\sigma_y\otimes P_1+I\sigma_y\otimes F_1\otimes P_2+\sigma_x\otimes F_2\otimes P_2$
$-A_{011}-B_{010}+C_{001}+D_{000}+E_{111}-F_{110}+G_{101}+H_{100}$	$\sigma_x\otimes I\sigma_y\otimes P_1-I\sigma_y\otimes F_1\otimes P_2+\sigma_x\otimes F_2\otimes P_2$
$-A_{011}+B_{010}+C_{001}+D_{000}-E_{111}-F_{110}+G_{101}+H_{100}$	$I\sigma_y\otimes F_1\otimes P_1+\sigma_x\otimes F_2\otimes P_1+\sigma_x\otimes I\sigma_y\otimes P_2$
$A_{011}-B_{010}+C_{001}+D_{000}-E_{111}-F_{110}+G_{101}+H_{100}$	$-I\sigma_y\otimes F_1\otimes P_1+\sigma_x\otimes F_2\otimes P_1+\sigma_x\otimes I\sigma_y\otimes P_2$
$-A_{011}-B_{010}+C_{001}+D_{000}-E_{111}-F_{110}+G_{101}+H_{100}$	$\sigma_x\otimes I\sigma_y\otimes P_1+\sigma_x\otimes I\sigma_y\otimes P_2$
$A_{011}+B_{010}+C_{001}+D_{000}+E_{111}+F_{110}+G_{101}-H_{100}$	$\sigma_x\otimes\sigma_x\otimes P_1+\sigma_x\otimes F_1\otimes P_2-I\sigma_y\otimes F_2\otimes P_2$
$A_{011}+B_{010}+C_{001}+D_{000}-E_{111}+F_{110}+G_{101}-H_{100}$	$\sigma_x\otimes\sigma_x\otimes P_1+\sigma_x\otimes I\sigma_y\otimes P_2$
$A_{011}+B_{010}+C_{001}+D_{000}+E_{111}-F_{110}+G_{101}-H_{100}$	$\sigma_x\otimes\sigma_x\otimes P_1-I\sigma_y\otimes\sigma_x\otimes P_2$
$A_{011}+B_{010}+C_{001}+D_{000}+E_{111}-F_{110}-G_{101}+H_{100}$	$\sigma_x\otimes\sigma_x\otimes P_1+I\sigma_y\otimes I\sigma_y\otimes P_2$
$A_{011}+B_{010}+C_{001}+D_{000}+E_{111}+F_{110}-G_{101}-H_{100}$	$\sigma_x\otimes\sigma_x\otimes P_1-\sigma_x\otimes I\sigma_y\otimes P_2$
$A_{011}+B_{010}+C_{001}+D_{000}-E_{111}-F_{110}-G_{101}+H_{100}$	$\sigma_x\otimes\sigma_x\otimes P_1-\sigma_x\otimes F_1\otimes P_2+I\sigma_y\otimes F_2\otimes P_2$
$A_{011}+B_{010}+C_{001}+D_{000}-E_{111}-F_{110}+G_{101}-H_{100}$	$\sigma_x\otimes\sigma_x\otimes P_1-\sigma_x\otimes F_1\otimes P_2-I\sigma_y\otimes F_2\otimes P_2$
$A_{011}+B_{010}+C_{001}+D_{000}-E_{111}+F_{110}-G_{101}-H_{100}$	$\sigma_x\otimes\sigma_x\otimes P_1+I\sigma_y\otimes F_1\otimes P_2-\sigma_x\otimes F_2\otimes P_2$
$A_{011}+B_{010}+C_{001}+D_{000}+E_{111}-F_{110}-G_{101}-H_{100}$	$\sigma_x\otimes\sigma_x\otimes P_1-I\sigma_y\otimes F_1\otimes P_2-\sigma_x\otimes F_2\otimes P_2$
$-A_{100}-B_{101}-C_{110}-D_{111}+E_{000}+F_{001}+G_{010}+H_{011}$	$-I_2\otimes I_2\otimes F_1+I_2\otimes I_2\otimes F_2$
$A_{100}+B_{101}+C_{110}+D_{111}+E_{000}+F_{001}+G_{010}+H_{011}$	$I_2\otimes I_2\otimes F_1+I_2\otimes I_2\otimes F_2$
$A_{100}-B_{101}+C_{110}+D_{111}+E_{000}+F_{001}+G_{010}+H_{011}$	$\sigma_z\otimes P_1\otimes F_1+I_2\otimes P_2\otimes F_1+I_2\otimes I_2\otimes F_2$
$A_{100}+B_{101}+C_{110}-D_{111}+E_{000}+F_{001}+G_{010}+H_{011}$	$I_2\otimes P_1\otimes F_1+\sigma_z\otimes P_2\otimes F_1+I_2\otimes I_2\otimes F_2$
$A_{100}+B_{101}+C_{110}+D_{111}+E_{000}-F_{001}+G_{010}+H_{011}$	$I_2\otimes I_2\otimes F_1+\sigma_z\otimes P_1\otimes F_2+I_2\otimes P_2\otimes F_2$
$A_{100}+B_{101}+C_{110}+D_{111}+E_{000}+F_{001}+G_{010}-H_{011}$	$I_2\otimes I_2\otimes F_1+I_2\otimes P_1\otimes F_2+\sigma_z\otimes P_2\otimes F_2$
$A_{100}-B_{101}+C_{110}-D_{111}+E_{000}+F_{001}+G_{010}+H_{011}$	$\sigma_z\otimes I_2\otimes F_1+I_2\otimes I_2\otimes F_2$
$A_{100}-B_{101}+C_{110}+D_{111}+E_{000}-F_{001}+G_{010}+H_{011}$	$\sigma_z\otimes P_1\otimes F_1+I_2\otimes P_2\otimes F_1+\sigma_z\otimes P_1\otimes F_2+I_2\otimes P_2\otimes F_2$
$A_{100}-B_{101}+C_{110}+D_{111}+E_{000}+F_{001}+G_{010}-H_{011}$	$\sigma_z\otimes P_1\otimes F_1+I_2\otimes P_2\otimes F_1+I_2\otimes P_1\otimes F_2+\sigma_z\otimes P_2\otimes F_2$
$A_{100}+B_{101}+C_{110}-D_{111}+E_{000}-F_{001}+G_{010}+H_{011}$	$I_2\otimes P_1\otimes F_1+\sigma_z\otimes P_2\otimes F_1+\sigma_z\otimes P_1\otimes F_2+I_2\otimes P_2\otimes F_2$
$A_{100}+B_{101}+C_{110}-D_{111}+E_{000}+F_{001}+G_{010}-H_{011}$	$I_2\otimes P_1\otimes F_1+\sigma_z\otimes P_2\otimes F_1+I_2\otimes P_1\otimes F_2+\sigma_z\otimes P_2\otimes F_2$
$A_{100}+B_{101}+C_{110}+D_{111}+E_{000}-F_{001}+G_{010}-H_{011}$	$I_2\otimes I_2\otimes F_1+\sigma_z\otimes I_2\otimes F_2$
$A_{100}-B_{101}+C_{110}-D_{111}+E_{000}-F_{001}+G_{010}+H_{011}$	$\sigma_z\otimes I_2\otimes F_1+\sigma_z\otimes P_1\otimes F_2+I_2\otimes P_2\otimes F_2$
$A_{100}-B_{101}+C_{110}+D_{111}+E_{000}-F_{001}+G_{010}-H_{011}$	$\sigma_z\otimes P_1\otimes F_1+I_2\otimes P_2\otimes F_1+\sigma_z\otimes I_2\otimes F_2$
$A_{100}-B_{101}+C_{110}-D_{111}+E_{000}+F_{001}+G_{010}-H_{011}$	$\sigma_z\otimes I_2\otimes F_1+I_2\otimes P_1\otimes F_2+\sigma_z\otimes P_2\otimes\sigma_z$
$A_{100}+B_{101}+C_{110}-D_{111}+E_{000}-F_{001}+G_{010}-H_{011}$	$I_2\otimes P_1\otimes F_1+\sigma_z\otimes P_2\otimes F_1+\sigma_z\otimes I_2\otimes F_2$
$A_{100}-B_{101}+C_{110}-D_{111}+E_{000}-F_{001}+G_{010}-H_{011}$	$\sigma_z\otimes I_2\otimes F_1+\sigma_z\otimes I_2\otimes F_2$
$A_{100}+B_{101}-C_{110}+D_{111}+E_{000}+F_{001}+G_{010}+H_{011}$	$I_2\otimes P_1\otimes F_1-\sigma_z\otimes P_2\otimes F_1+I_2\otimes I_2\otimes F_2$
$A_{100}+B_{101}+C_{110}+D_{111}+E_{000}+F_{001}-G_{010}+H_{011}$	$I_2\otimes I_2\otimes F_1+I_2\otimes P_1\otimes F_2-\sigma_z\otimes P_2\otimes F_2$
$A_{100}+B_{101}-C_{110}-D_{111}+E_{000}+F_{001}+G_{010}+H_{011}$	$I_2\otimes\sigma_z\otimes F_1+I_2\otimes I_2\otimes F_2$
$A_{100}+B_{101}-C_{110}+D_{111}+E_{000}+F_{001}-G_{010}+H_{011}$	$I_2\otimes P_1\otimes F_1-\sigma_z\otimes P_2\otimes F_1+I_2\otimes P_1\otimes F_2-\sigma_z\otimes P_2\otimes F_2$

State Obtained by Alice	Short-Hand Form of Transformation
$A_{100} + B_{101} - C_{110} + D_{111} + E_{000} + F_{001} + G_{010} - H_{011}$	$I_2 \otimes P_1 \otimes - \sigma_z \otimes P_2 \otimes F_1 + I_2 \otimes P_1 \otimes F_2 + \sigma_z \otimes P_2 \otimes F_2$
$A_{100} + B_{101} + C_{110} - D_{111} + E_{000} + F_{001} - G_{010} + H_{011}$	$I_2 \otimes P_1 \otimes F_1 + \sigma_z \otimes P_2 \otimes F_1 + I_2 \otimes P_1 \otimes F_2 - \sigma_z \otimes P_2 \otimes F_2$
$A_{100} + B_{101} + C_{110} + D_{111} + E_{000} + F_{001} - G_{010} - H_{011}$	$I_2 \otimes I_2 \otimes F_1 + I_2 \otimes \sigma_z \otimes F_2$
$A_{100} + B_{101} - C_{110} - D_{111} + E_{000} + F_{001} - G_{010} + H_{011}$	$I_2 \otimes \sigma_z \otimes F_1 + I_2 \otimes I_2 \otimes F_2$
$A_{100} + B_{101} - C_{110} - D_{111} + E_{000} + F_{001} + G_{010} - H_{011}$	$I_2 \otimes \sigma_z \otimes F_1 + I_2 \otimes P_1 \otimes F_2 + \sigma_z \otimes P_2 \otimes F_2$
$A_{100} + B_{101} - C_{110} + D_{111} + E_{000} + F_{001} - G_{010} - H_{011}$	$I_2 \otimes P_1 \otimes F_1 - \sigma_z \otimes P_2 \otimes F_1 + I_2 \otimes \sigma_z \otimes F_2$
$A_{100} + B_{101} + C_{110} - D_{111} + E_{000} + F_{001} - G_{010} - H_{011}$	$I_2 \otimes P_1 \otimes F_1 + \sigma_z \otimes P_2 \otimes F_1 + I_2 \otimes \sigma_z \otimes F_2$
$A_{100} + B_{101} - C_{110} - D_{111} + E_{000} + F_{001} - G_{010} - H_{011}$	$I_2 \otimes \sigma_z \otimes F_1 + I_2 \otimes \sigma_z \otimes F_2$
$-A_{100} + B_{101} + C_{110} + D_{111} + E_{000} + F_{001} + G_{010} + H_{011}$	$-\sigma_z \otimes P_1 \otimes F_1 + I_2 \otimes P_2 \otimes F_1 + I_2 \otimes I_2 \otimes F_2$
$-A_{100} - B_{101} + C_{110} + D_{111} + E_{000} + F_{001} + G_{010} + H_{011}$	$-I_2 \otimes \sigma_z \otimes F_1 + I_2 \otimes I_2 \otimes F_2$
$-A_{100} + B_{101} - C_{110} + D_{111} + E_{000} + F_{001} + G_{010} + H_{011}$	$-\sigma_z \otimes I_2 \otimes F_1 + I_1 \otimes I_2 \otimes F_2$
$-A_{100} + B_{101} + C_{110} - D_{111} + E_{000} + F_{001} + G_{010} + H_{011}$	$-\sigma_z \otimes \sigma_z \otimes F_1 + I_2 \otimes I_2 \otimes F_2$
$A_{100} - B_{101} - C_{110} + D_{111} + E_{000} + F_{001} + G_{010} + H_{011}$	$\sigma_z \otimes \sigma_z \otimes F_1 + I_2 \otimes I_2 \otimes F_2$
$-A_{100} - B_{101} - C_{110} + D_{111} + E_{000} + F_{001} + G_{010} + H_{011}$	$-I_2 \otimes P_1 \otimes F_1 - \sigma_z \otimes P_2 \otimes F_1 + I_2 \otimes I_2 \otimes F_2$
$-A_{100} - B_{101} + C_{110} - D_{111} + E_{000} + F_{001} + G_{010} + H_{011}$	$-I_2 \otimes P_1 \otimes F_1 + \sigma_z \otimes P_2 \otimes F_1 + I_2 \otimes I_2 \otimes F_2$
$-A_{100} + B_{101} - C_{110} - D_{111} + E_{000} + F_{001} + G_{010} + H_{011}$	$-\sigma_z \otimes P_1 \otimes F_1 - I_2 \otimes P_2 \otimes F_1 + I_2 \otimes I_2 \otimes F_2$
$A_{100} - B_{101} - C_{110} - D_{111} + E_{000} + F_{001} + G_{010} + H_{011}$	$\sigma_z \otimes P_1 \otimes F_1 - I_2 \otimes P_2 \otimes F_1 + I_2 \otimes P_1 \otimes F_2 + I_2 \otimes P_2 \otimes F_2$
$A_{101} + B_{100} + C_{111} + D_{110} + E_{001} + F_{000} + G_{011} + H_{010}$	$\sigma_x \otimes I_2 \otimes F_1 + \sigma_x \otimes I_2 \otimes F_2$
$-A_{101} + B_{100} + C_{111} + D_{110} + E_{001} + F_{000} + G_{011} + H_{010}$	$I\sigma_y \otimes P_1 \otimes F_1 + \sigma_x \otimes P_2 \otimes F_1 + \sigma_x \otimes I_2 \otimes F_2$
$A_{101} + B_{100} - C_{111} + D_{110} + E_{001} + F_{000} + G_{011} + H_{010}$	$\sigma_x \otimes P_1 \otimes F_1 + I\sigma_y \otimes P_2 \otimes F_1 + \sigma_x \otimes I_2 \otimes F_2$
$A_{101} + B_{100} + C_{111} + D_{110} - E_{001} + F_{000} + G_{011} + H_{010}$	$\sigma_x \otimes I_2 \otimes F_1 + I\sigma_y \otimes P_1 \otimes F_2 + \sigma_x \otimes P_2 \otimes F_2$
$A_{101} + B_{100} + C_{111} + D_{110} + E_{001} + F_{000} - G_{011} + H_{010}$	$\sigma_x \otimes I_2 \otimes F_1 + \sigma_x \otimes P_1 \otimes F_2 + I\sigma_y \otimes P_2 \otimes F_2$
$-A_{101} + B_{100} - C_{111} + D_{110} + E_{001} + F_{000} + G_{011} + H_{010}$	$I\sigma_y \otimes I_2 \otimes F_1 + \sigma_x \otimes I_2 \otimes F_2$
$-A_{101} + B_{100} + C_{111} + D_{110} - E_{001} + F_{000} + G_{011} + H_{010}$	$I\sigma_y \otimes P_1 \otimes F_1 + \sigma_x \otimes P_2 \otimes F_1 + I\sigma_y \otimes P_1 \otimes F_2 + \sigma_x \otimes P_2 \otimes F_2$
$-A_{101} + B_{100} + C_{111} + D_{110} + E_{001} + F_{000} - G_{011} + H_{010}$	$I\sigma_y \otimes P_1 \otimes F_1 + \sigma_x \otimes P_2 \otimes F_1 + \sigma_x \otimes P_1 \otimes F_2 + I\sigma_y \otimes P_2 \otimes F_2$
$A_{101} + B_{100} - C_{111} + D_{110} - E_{001} + F_{000} + G_{011} + H_{010}$	$\sigma_x \otimes P_1 \otimes F_1 + I\sigma_y \otimes P_2 \otimes F_1 + I\sigma_y \otimes P_1 \otimes F_2 + \sigma_x \otimes P_2 \otimes F_2$
$A_{101} + B_{100} - C_{111} + D_{110} + E_{001} + F_{000} - G_{011} + H_{010}$	$\sigma_x \otimes P_1 \otimes F_1 + I\sigma_y \otimes P_2 \otimes F_1 + \sigma_x \otimes P_1 \otimes F_2 + I\sigma_y \otimes P_2 \otimes F_2$
$A_{101} + B_{100} + C_{111} + D_{110} - E_{001} + F_{000} - G_{011} + H_{010}$	$\sigma_x \otimes I_2 \otimes F_1 + I\sigma_y \otimes I_2 \otimes F_2$
$-A_{101} + B_{100} - C_{111} + D_{110} - E_{001} + F_{000} + G_{011} + H_{010}$	$I\sigma_y \otimes I_2 \otimes F_1 + I\sigma_y \otimes P_1 \otimes F_2 + \sigma_x \otimes P_2 \otimes F_2$
$-A_{101} + B_{100} + C_{111} + D_{110} - E_{001} + F_{000} - G_{011} + H_{010}$	$I\sigma_y \otimes P_1 \otimes F_1 + \sigma_x \otimes P_2 \otimes F_1 + I\sigma_y \otimes I_2 \otimes F_2$
$-A_{101} + B_{100} - C_{111} + D_{110} + E_{001} + F_{000} - G_{011} + H_{010}$	$I\sigma_y \otimes I_2 \otimes F_1 + \sigma_x \otimes P_1 \otimes F_2 + I\sigma_y \otimes P_2 \otimes F_2$
$A_{101} + B_{100} - C_{111} + D_{110} - E_{001} + F_{000} - G_{011} + H_{010}$	$\sigma_x \otimes P_1 \otimes F_1 + I\sigma_y \otimes P_2 \otimes F_1 + I\sigma_y \otimes P_1 \otimes F_2 + I\sigma_y \otimes P_2 \otimes F_2$
$-A_{101} + B_{100} - C_{111} + D_{110} - E_{001} + F_{000} - G_{011} + H_{010}$	$I\sigma_y \otimes I_2 \otimes F_1 + I\sigma_y \otimes I_2 \otimes F_2$
$A_{101} + B_{100} + C_{111} - D_{110} + E_{001} + F_{000} + G_{011} - H_{010}$	$\sigma_x \otimes P_1 \otimes F_1 - I\sigma_y \otimes P_2 \otimes F_1 + \sigma_x \otimes P_1 \otimes F_2 - I\sigma_y \otimes P_2 \otimes F_2$
$A_{101} + B_{100} + C_{111} - D_{110} + E_{001} + F_{000} + G_{011} + H_{010}$	$\sigma_x \otimes P_1 \otimes F_1 - I\sigma_y \otimes P_2 \otimes F_1 + \sigma_x \otimes I_2 \otimes F_2$
$A_{101} + B_{100} + C_{111} + D_{110} + E_{001} + F_{000} + G_{011} - H_{010}$	$\sigma_x \otimes I_2 \otimes F_1 + \sigma_x \otimes P_1 \otimes F_2 - I\sigma_y \otimes P_2 \otimes F_2$
$A_{101} + B_{100} - C_{111} - D_{110} + E_{001} + F_{000} + G_{011} + H_{010}$	$\sigma_x \otimes \sigma_z \otimes F_1 + \sigma_x \otimes I_2 \otimes F_2$
$A_{101} + B_{100} - C_{111} + D_{110} + E_{001} + F_{000} + G_{011} - H_{010}$	$\sigma_x \otimes P_1 \otimes F_1 + I\sigma_y \otimes P_2 \otimes F_1 + \sigma_x \otimes P_1 \otimes F_2 - I\sigma_y \otimes P_2 \otimes F_2$
$A_{101} + B_{100} + C_{111} - D_{110} + E_{001} + F_{000} - G_{011} + H_{010}$	$I\sigma_y \otimes P_1 \otimes F_1 + \sigma_x \otimes P_2 \otimes F_1 + \sigma_x \otimes I_2 \otimes F_2$
$A_{101} + B_{100} + C_{111} + D_{110} + E_{001} + F_{000} - G_{011} - H_{010}$	$\sigma_x \otimes I_2 \otimes F_1 + \sigma_x \otimes \sigma_z \otimes F_2$
$A_{101} + B_{100} - C_{111} - D_{110} + E_{001} + F_{000} - G_{011} + H_{010}$	$\sigma_x \otimes \sigma_z \otimes F_1 + \sigma_x \otimes P_1 \otimes F_2 + I\sigma_y \otimes P_2 \otimes F_2$
$A_{101} + B_{100} - C_{111} - D_{110} + E_{001} + F_{000} + G_{011} - H_{010}$	$\sigma_x \otimes \sigma_z \otimes F_1 + \sigma_x \otimes P_1 \otimes F_2 - I\sigma_y \otimes P_2 \otimes F_2$
$A_{101} + B_{100} - C_{111} + D_{110} + E_{001} + F_{000} - G_{011} - H_{010}$	$\sigma_x \otimes P_1 \otimes F_1 + I\sigma_y \otimes P_2 \otimes F_1 + \sigma_x \otimes \sigma_z \otimes F_2$

State Obtained by Alice	Short-Hand Form of Transformation
$A_{101} + B_{100} + C_{111} - D_{110} + E_{001} + F_{000} - G_{011} - H_{010}$	$\sigma_x \otimes P_1 \otimes F_1 - I\sigma_y \otimes P_2 \otimes F_1 + \sigma_x \otimes \sigma_z \otimes F_2$
$A_{101} + B_{100} - C_{111} - D_{110} + E_{001} + F_{000} - G_{011} - H_{010}$	$\sigma_x \otimes \sigma_z \otimes F_1 + \sigma_x \otimes \sigma_z \otimes F_2$
$A_{101} - B_{100} + C_{111} + D_{110} + E_{001} + F_{000} + G_{011} + H_{010}$	$-I\sigma_y \otimes P_1 \otimes F_1 + \sigma_x \otimes P_2 \otimes F_1 + \sigma_x \otimes I_2 \otimes F_2$
$-A_{101} - B_{100} + C_{111} + D_{110} + E_{001} + F_{000} + G_{011} + H_{010}$	$-\sigma_x \otimes \sigma_z \otimes F_1 + \sigma_x \otimes I_2 \otimes F_2$
$-A_{101} + B_{100} + C_{111} - D_{110} + E_{001} + F_{000} + G_{011} + H_{010}$	$I\sigma_y \otimes P_1 \otimes F_1 - I\sigma_y \otimes P_2 \otimes F_1 + I\sigma_y \otimes I_2 \otimes F_2$
$A_{101} - B_{100} - C_{111} + D_{110} + E_{001} + F_{000} + G_{011} + H_{010}$	$-I\sigma_y \otimes \sigma_z \otimes F_1 + \sigma_x \otimes I_2 \otimes F_2$
$A_{101} - B_{100} + C_{111} - D_{110} + E_{001} + F_{000} + G_{011} + H_{010}$	$-I\sigma_y \otimes I_2 \otimes F_1 + \sigma_x \otimes I_2 \otimes F_2$
$-A_{101} - B_{100} - C_{111} + D_{110} + E_{001} + F_{000} + G_{011} + H_{010}$	$-\sigma_x \otimes P_1 \otimes F_1 + I\sigma_y \otimes P_2 \otimes F_1 + I\sigma_y \otimes I_2 \otimes F_2$
$-A_{101} - B_{100} + C_{111} - D_{110} + E_{001} + F_{000} + G_{011} + H_{010}$	$-F_1 \otimes P_1 \otimes F_1 - I\sigma_y \otimes P_2 \otimes F_1 + I\sigma_y \otimes I_2 \otimes F_2$
$-A_{101} + B_{100} - C_{111} - D_{110} + E_{001} + F_{000} + G_{011} + H_{010}$	$I\sigma_y \otimes I_2 \otimes F_1 + \sigma_x \otimes I_2 \otimes F_2$
$A_{101} - B_{100} - C_{111} - D_{110} + E_{001} + F_{000} + G_{011} + H_{010}$	$-I\sigma_y \otimes P_1 \otimes F_1 - \sigma_x \otimes P_2 \otimes F_1 + \sigma_x \otimes I_2 \otimes F_2$
$-A_{101} - B_{100} - C_{111} - D_{110} + E_{001} + F_{000} + G_{011} + H_{010}$	$-\sigma_x \otimes I_2 \otimes F_1 + \sigma_x \otimes I_2 \otimes F_2$
$A_{110} - B_{111} + C_{100} + D_{101} + E_{010} + F_{000} + G_{000} + H_{001}$	$\sigma_z \otimes F_1 \otimes F_1 + I_2 \otimes F_2 \otimes F_1 + I_2 \otimes \sigma_x \otimes F_2$
$A_{110} + B_{111} + C_{100} + D_{101} + E_{010} + F_{000} + G_{000} + H_{001}$	$I_2 \otimes \sigma_x \otimes F_1 + I_2 \otimes \sigma_x \otimes F_2$
$+A_{110} + B_{111} + C_{100} - D_{101} + E_{010} + F_{000} + G_{000} + H_{001}$	$I_2 \otimes F_1 \otimes F_1 + \sigma_z \otimes F_2 \otimes F_1 + I_2 \otimes \sigma_x \otimes F_2$
$A_{110} + B_{111} + C_{100} + D_{101} + E_{010} + F_{000} + G_{000} - H_{001}$	$I_2 \otimes \sigma_x \otimes F_1 + I_2 \otimes \sigma_x \otimes F_2 + \sigma_z \otimes F_2 \otimes F_1$
$A_{110} + B_{111} + C_{100} + D_{101} + E_{010} - F_{000} + G_{000} + H_{001}$	$I_2 \otimes \sigma_x \otimes F_1 + \sigma_z \otimes F_1 \otimes F_2 + I_2 \otimes F_2 \otimes F_2$
$A_{110} - B_{111} + C_{100} - D_{101} + E_{010} + F_{000} + G_{000} + H_{001}$	$\sigma_z \otimes \sigma_x \otimes F_1 + I_2 \otimes \sigma_x \otimes F_2$
$A_{110} - B_{111} + C_{100} + D_{101} + E_{010} - F_{000} + G_{000} + H_{001}$	$\sigma_z \otimes F_1 \otimes F_1 + I_2 \otimes F_1 \otimes F_2 + \sigma_z \otimes F_1 \otimes F_2 + I_2 \otimes F_2 \otimes F_2$
$A_{110} - B_{111} + C_{100} + D_{101} + E_{010} + F_{000} + G_{000} - H_{001}$	$\sigma_z \otimes F_1 \otimes F_1 + I_2 \otimes F_2 \otimes F_1 + I_2 \otimes F_1 \otimes F_2 + \sigma_z \otimes F_2 \otimes F_2$
$A_{110} + B_{111} + C_{100} - D_{101} + E_{010} - F_{000} + G_{000} + H_{001}$	$I_2 \otimes F_1 \otimes F_1 + \sigma_z \otimes F_2 \otimes F_1 + \sigma_z \otimes F_1 \otimes F_2 + I_2 \otimes F_2 \otimes F_2$
$A_{110} + B_{111} + C_{100} - D_{101} + E_{010} + F_{000} + G_{000} - H_{001}$	$I_2 \otimes F_1 \otimes F_1 + \sigma_z \otimes F_2 \otimes F_1 + I_2 \otimes F_1 \otimes F_2 + \sigma_z \otimes F_2 \otimes F_2$
$A_{110} + B_{111} + C_{100} + D_{101} + E_{010} - F_{000} + G_{000} - H_{001}$	$I_2 \otimes \sigma_x \otimes F_1 + \sigma_z \otimes \sigma_x \otimes F_2$
$A_{110} - B_{111} + C_{100} - D_{101} + E_{010} - F_{000} + G_{000} + H_{001}$	$\sigma_z \otimes \sigma_x \otimes F_1 + \sigma_z \otimes F_1 \otimes F_2 + I_2 \otimes F_2 \otimes F_2$
$A_{110} - B_{111} + C_{100} - D_{101} + E_{010} + F_{000} + G_{000} - H_{001}$	$\sigma_z \otimes \sigma_x \otimes F_1 + I_2 \otimes F_1 \otimes F_2 + \sigma_z \otimes F_2 \otimes F_2$
$A_{110} - B_{111} + C_{100} + D_{101} + E_{010} - F_{000} + G_{000} - H_{001}$	$\sigma_z \otimes F_1 \otimes F_1 + I_2 \otimes F_2 \otimes F_1 + \sigma_z \otimes F_1 \otimes F_2 + \sigma_z \otimes F_2 \otimes F_2$
$A_{110} + B_{111} + C_{100} - D_{101} + E_{010} - F_{000} + G_{000} - H_{001}$	$I_2 \otimes F_1 \otimes F_1 + \sigma_z \otimes F_2 \otimes F_1 + \sigma_z \otimes F_1 \otimes F_2 + \sigma_z \otimes F_2 \otimes F_2$
$A_{110} - B_{111} + C_{100} - D_{101} + E_{010} - F_{000} + G_{000} - H_{001}$	$\sigma_z \otimes \sigma_x \otimes F_1 + \sigma_z \otimes \sigma_x \otimes F_2$
$-A_{110} + B_{111} + C_{100} + D_{101} + E_{010} + F_{000} + G_{000} + H_{001}$	$-\sigma_z \otimes F_1 \otimes F_1 + I_2 \otimes F_2 \otimes F_1 + I_2 \otimes \sigma_x \otimes F_2$
$A_{110} + B_{111} + C_{100} + D_{101} - E_{010} + F_{000} + G_{000} + H_{001}$	$I_2 \otimes \sigma_x \otimes F_1 - \sigma_z \otimes F_1 \otimes F_2 + I_2 \otimes F_2 \otimes F_2$
$-A_{110} - B_{111} + C_{100} + D_{101} + E_{010} + F_{000} + G_{000} + H_{001}$	$I_2 \otimes I\sigma_y \otimes F_1 + I_2 \otimes \sigma_x \otimes F_2$
$-A_{110} + B_{111} + C_{100} + D_{101} - E_{010} + F_{000} + G_{000} + H_{001}$	$-\sigma_z \otimes F_1 \otimes F_1 + I_2 \otimes F_2 \otimes F_1 - \sigma_z \otimes F_1 \otimes F_2 + I_2 \otimes F_2 \otimes F_2$
$-A_{110} + B_{111} + C_{100} + D_{101} + E_{010} - F_{000} + G_{000} + H_{001}$	$-\sigma_z \otimes F_1 \otimes F_1 + I_2 \otimes F_2 \otimes F_1 + \sigma_z \otimes F_1 \otimes F_2 + I_2 \otimes F_2 \otimes F_2$
$A_{110} - B_{111} + C_{100} + D_{101} - E_{010} + F_{000} + G_{000} + H_{001}$	$\sigma_z \otimes F_1 \otimes F_1 + I_2 \otimes F_2 \otimes F_1 - \sigma_z \otimes F_1 \otimes F_2 + I_2 \otimes F_2 \otimes F_2$
$A_{110} + B_{111} + C_{100} + D_{101} - E_{010} - F_{000} + G_{000} + H_{001}$	$I_2 \otimes \sigma_x \otimes F_1 + I_2 \otimes I\sigma_y \otimes F_2$
$-A_{110} - B_{111} + C_{100} + D_{101} - E_{010} + F_{000} + G_{000} + H_{001}$	$\sigma_z \otimes F_1 \otimes F_1 + I_2 \otimes F_2 \otimes F_1 - \sigma_z \otimes F_1 \otimes F_2 + I_2 \otimes F_2 \otimes F_2$
$-A_{110} + B_{111} + C_{100} + D_{101} - E_{010} - F_{000} + G_{000} + H_{001}$	$-\sigma_z \otimes F_1 \otimes F_1 + I_2 \otimes F_2 \otimes F_1 + I_2 \otimes I\sigma_y \otimes F_2$
$-A_{110} - B_{111} + C_{100} + D_{101} + E_{010} - F_{000} + G_{000} + H_{001}$	$I_2 \otimes I\sigma_y \otimes F_1 + \sigma_z \otimes F_1 \otimes F_2 + I_2 \otimes F_2 \otimes F_2$
$A_{110} - B_{111} + C_{100} + D_{101} - E_{010} - F_{000} + G_{000} + H_{001}$	$\sigma_z \otimes F_1 \otimes F_1 + I_2 \otimes F_2 \otimes F_1 + I_2 \otimes I\sigma_y \otimes F_2$
$-A_{110} - B_{111} + C_{100} + D_{101} - E_{010} - F_{000} + G_{000} + H_{001}$	$I_2 \otimes I\sigma_y \otimes F_1 + I_2 \otimes I\sigma_y \otimes F_2$
$A_{110} + B_{111} - C_{100} + D_{101} + E_{010} + F_{000} + G_{000} + H_{001}$	$I_2 \otimes F_1 \otimes F_1 - \sigma_z \otimes F_2 \otimes F_1 + I_2 \otimes \sigma_x \otimes F_2$
$-A_{110} + B_{111} - C_{100} + D_{101} + E_{010} + F_{000} + G_{000} + H_{001}$	$-\sigma_z \otimes F_1 \otimes F_1 - \sigma_z \otimes F_2 \otimes F_1 + I_2 \otimes \sigma_x \otimes F_2$
$-A_{110} + B_{111} + C_{100} - D_{101} + E_{010} + F_{000} + G_{000} + H_{001}$	$\sigma_z \otimes I\sigma_y \otimes F_1 + I_2 \otimes \sigma_x \otimes F_2$

State Obtained by Alice	Short-Hand Form of Transformation
$A_{110} - B_{111} - C_{100} + D_{101} + E_{010} + F_{000} + G_{000} + H_{001}$	$-\sigma_z \otimes I\sigma_y \otimes F_1 + I_2 \otimes \sigma_x \otimes F_2$
$A_{110} + B_{111} - C_{100} - D_{101} + E_{010} + F_{000} + G_{000} + H_{001}$	$I_2 \otimes I\sigma_y \otimes F_1 + I_2 \otimes \sigma_x \otimes F_2$
$-A_{110} - B_{111} - C_{100} + D_{101} + E_{010} + F_{000} + G_{000} + H_{001}$	$I_2 \otimes F_1 \otimes F_1 - \sigma_z \otimes F_2 \otimes F_1 + I_2 \otimes \sigma_x \otimes F_2$
$-A_{110} - B_{111} + C_{100} - D_{101} + E_{010} + F_{000} + G_{000} + H_{001}$	$-I_2 \otimes F_1 \otimes F_1 + \sigma_z \otimes F_2 \otimes F_1 + I_2 \otimes \sigma_x \otimes F_2$
$-A_{110} + B_{111} - C_{100} - D_{101} + E_{010} + F_{000} + G_{000} + H_{001}$	$-\sigma_z \otimes F_1 \otimes F_1 - I_2 \otimes F_2 \otimes F_1 + I_2 \otimes \sigma_x \otimes F_2$
$A_{110} - B_{111} - C_{100} - D_{101} + E_{010} + F_{000} + G_{000} + H_{001}$	$\sigma_z \otimes F_1 \otimes F_1 - I_2 \otimes F_2 \otimes F_1 + I_2 \otimes \sigma_x \otimes F_2$
$-A_{110} - B_{111} - C_{100} - D_{101} + E_{010} + F_{000} + G_{000} + H_{001}$	$-I_2 \otimes \sigma_x \otimes F_1 + I_2 \otimes \sigma_x \otimes F_2$
$-A_{111} + B_{110} + C_{101} + D_{100} + E_{011} + F_{010} + G_{001} + H_{000}$	$I\sigma_y \otimes F_1 \otimes F_1 + \sigma_x \otimes F_2 \otimes F_1 + \sigma_x \otimes \sigma_x \otimes F_2$
$A_{111} + B_{110} - C_{101} + D_{100} + E_{011} + F_{010} + G_{001} + H_{000}$	$\sigma_x \otimes F_1 \otimes F_1 + I\sigma_y \otimes F_2 \otimes F_1 + \sigma_x \otimes \sigma_x \otimes F_2$
$A_{111} + B_{110} + C_{101} + D_{100} - E_{011} + F_{010} + G_{001} + H_{000}$	$\sigma_x \otimes \sigma_x \otimes F_1 + I\sigma_y \otimes F_1 \otimes F_2 + \sigma_x \otimes F_2 \otimes F_2$
$A_{111} + B_{110} + C_{101} + D_{100} + E_{011} + F_{010} - G_{001} + H_{000}$	$\sigma_x \otimes \sigma_x \otimes F_1 + \sigma_x \otimes F_1 \otimes F_2 + I\sigma_y \otimes F_2 \otimes F_2$
$-A_{111} + B_{110} - C_{101} + D_{100} + E_{011} + F_{010} + G_{001} + H_{000}$	$I\sigma_y \otimes \sigma_x \otimes F_1 + \sigma_x \otimes \sigma_x \otimes F_2$
$-A_{111} + B_{110} + C_{101} + D_{100} - E_{011} + F_{010} + G_{001} + H_{000}$	$I\sigma_y \otimes F_1 \otimes F_1 + \sigma_x \otimes F_2 \otimes F_1 + I\sigma_y \otimes F_1 \otimes F_2 + \sigma_x \otimes F_2 \otimes F_2$
$-A_{111} + B_{110} + C_{101} + D_{100} + E_{011} + F_{010} - G_{001} + H_{000}$	$I\sigma_y \otimes F_1 \otimes F_1 + \sigma_x \otimes F_2 \otimes F_1 + I\sigma_y \otimes F_1 \otimes F_2 + \sigma_x \otimes F_2 \otimes F_2$
$A_{111} + B_{110} - C_{101} + D_{100} - E_{011} + F_{010} + G_{001} + H_{000}$	$\sigma_x \otimes F_1 \otimes F_1 + I\sigma_y \otimes F_2 \otimes F_1 + I\sigma_y \otimes F_1 \otimes F_2 + \sigma_x \otimes F_2 \otimes F_2$
$A_{111} + B_{110} - C_{101} + D_{100} + E_{011} + F_{010} - G_{001} + H_{000}$	$\sigma_x \otimes F_1 \otimes F_1 + I\sigma_y \otimes F_2 \otimes F_1 + \sigma_x \otimes F_1 \otimes F_2 + I\sigma_y \otimes F_2 \otimes F_2$
$A_{111} + B_{110} + C_{101} + D_{100} - E_{011} + F_{010} - G_{001} + H_{000}$	$\sigma_x \otimes \sigma_x \otimes F_1 + \sigma_x \otimes F_2 \otimes F_1 + I\sigma_y \otimes \sigma_x \otimes F_2$
$-A_{111} + B_{110} - C_{101} + D_{100} - E_{011} + F_{010} + G_{001} + H_{000}$	$I\sigma_y \otimes \sigma_x \otimes F_1 + I\sigma_y \otimes F_1 \otimes F_2 + \sigma_x \otimes F_2 \otimes F_2$
$-A_{111} + B_{110} + C_{101} + D_{100} - E_{011} + F_{010} - G_{001} + H_{000}$	$I\sigma_y \otimes F_1 \otimes F_1 + \sigma_x \otimes F_2 \otimes F_1 + I\sigma_y \otimes \sigma_x \otimes F_2$
$-A_{111} + B_{110} - C_{101} + D_{100} + E_{011} + F_{010} - G_{001} + H_{000}$	$I\sigma_y \otimes \sigma_x \otimes F_1 + \sigma_x \otimes F_1 \otimes F_2 + I\sigma_y \otimes F_2 \otimes F_2$
$A_{111} + B_{110} - C_{101} + D_{100} - E_{011} + F_{010} - G_{001} + H_{000}$	$\sigma_x \otimes F_1 \otimes F_1 + I\sigma_y \otimes F_2 \otimes F_1 + I\sigma_y \otimes \sigma_x \otimes F_2$
$-A_{111} + B_{110} - C_{101} + D_{100} - E_{011} + F_{010} - G_{001} + H_{000}$	$I\sigma_y \otimes \sigma_x \otimes F_1 + I\sigma_y \otimes \sigma_x \otimes F_2$
$A_{111} + B_{110} + C_{101} + D_{100} + E_{011} + F_{010} + G_{001} + H_{000}$	$\sigma_x \otimes \sigma_x \otimes \sigma_x$
$A_{111} - B_{110} + C_{101} + D_{100} + E_{011} + F_{010} + G_{001} + H_{000}$	$I\sigma_y \otimes F_1 \otimes F_1 + \sigma_x \otimes F_2 \otimes F_1 + \sigma_x \otimes \sigma_x \otimes F_2$
$A_{111} + B_{110} + C_{101} + D_{100} + E_{011} - F_{010} + G_{001} + H_{000}$	$\sigma_x \otimes \sigma_x \otimes F_1 - I\sigma_y \otimes F_1 \otimes F_2 + \sigma_x \otimes F_2 \otimes F_2$
$-A_{111} - B_{110} + C_{101} + D_{100} + E_{011} + F_{010} + G_{001} + H_{000}$	$\sigma_x \otimes I\sigma_y \otimes F_1 + \sigma_x \otimes \sigma_x \otimes F_2$
$-A_{111} + B_{110} + C_{101} + D_{100} + E_{011} - F_{010} + G_{001} + H_{000}$	$I\sigma_y \otimes F_1 \otimes F_1 + \sigma_x \otimes F_2 \otimes F_1 - I\sigma_y \otimes F_1 \otimes F_2 + \sigma_x \otimes F_2 \otimes F_2$
$A_{111} - B_{110} + C_{101} + D_{100} - E_{011} + F_{010} + G_{001} + H_{000}$	$-I\sigma_y \otimes F_1 \otimes F_1 + \sigma_x \otimes F_2 \otimes F_1 + I\sigma_y \otimes F_1 \otimes F_2 + \sigma_x \otimes F_2 \otimes F_2$
$A_{111} - B_{110} + C_{101} + D_{100} + E_{011} - F_{010} + G_{001} + H_{000}$	$-I\sigma_y \otimes F_1 \otimes F_1 + \sigma_x \otimes F_2 \otimes F_1 - I\sigma_y \otimes F_1 \otimes F_2 + \sigma_x \otimes F_2 \otimes F_2$
$A_{111} + B_{110} + C_{101} + D_{100} - E_{011} - F_{010} + G_{001} + H_{000}$	$\sigma_x \otimes \sigma_x \otimes F_1 + \sigma_x \otimes I\sigma_y \otimes F_2$
$-A_{111} - B_{110} + C_{101} + D_{100} - E_{011} + F_{010} + G_{001} + H_{000}$	$\sigma_x \otimes I\sigma_y \otimes F_1 + I\sigma_y \otimes F_1 \otimes F_2 + \sigma_x \otimes F_2 \otimes F_2$
$-A_{111} + B_{110} + C_{101} + D_{100} - E_{011} - F_{010} + G_{001} + H_{000}$	$I\sigma_y \otimes F_1 \otimes F_1 + \sigma_x \otimes F_2 \otimes F_1 + \sigma_x \otimes \sigma_x \otimes F_2$
$-A_{111} - B_{110} + C_{101} + D_{100} - E_{011} + F_{010} + G_{001} + H_{000}$	$\sigma_x \otimes I\sigma_y \otimes F_1 + I\sigma_y \otimes F_1 \otimes F_2 + \sigma_x \otimes F_2 \otimes F_2$
$A_{111} - B_{110} + C_{101} + D_{100} - E_{011} - F_{010} + G_{001} + H_{000}$	$-I\sigma_y \otimes F_1 \otimes F_1 + \sigma_x \otimes F_2 \otimes F_1 + \sigma_x \otimes I\sigma_y \otimes F_2$
$-A_{111} - B_{110} + C_{101} + D_{100} - E_{011} - F_{010} + G_{001} + H_{000}$	$\sigma_x \otimes I\sigma_y \otimes \sigma_x$
$A_{111} + B_{110} + C_{101} - D_{100} + E_{011} + F_{010} + G_{001} + H_{000}$	$\sigma_x \otimes F_1 \otimes F_1 - I\sigma_y \otimes F_2 \otimes F_1 + \sigma_x \otimes \sigma_x \otimes F_2$
$-A_{111} + B_{110} + C_{101} - D_{100} + E_{011} + F_{010} + G_{001} + H_{000}$	$-I\sigma_y \otimes I\sigma_y \otimes F_1 + \sigma_x \otimes \sigma_x \otimes F_2$
$A_{111} - B_{110} - C_{101} + D_{100} + E_{011} + F_{010} + G_{001} + H_{000}$	$I\sigma_y \otimes I\sigma_y \otimes F_1 + \sigma_x \otimes \sigma_x \otimes F_2$
$A_{111} - B_{110} + C_{101} - D_{100} + E_{011} + F_{010} + G_{001} + H_{000}$	$-I\sigma_y \otimes \sigma_x \otimes F_1 + \sigma_x \otimes \sigma_x \otimes F_2$
$A_{111} + B_{110} - C_{101} - D_{100} + E_{011} + F_{010} + G_{001} + H_{000}$	$-\sigma_x \otimes I\sigma_y \otimes F_1 + \sigma_x \otimes \sigma_x \otimes F_2$
$-A_{111} - B_{110} - C_{101} + D_{100} + E_{011} + F_{010} + G_{001} + H_{000}$	$-\sigma_x \otimes F_1 \otimes F_1 + I\sigma_y \otimes F_2 \otimes F_1 + \sigma_x \otimes \sigma_x \otimes F_2$
$-A_{111} - B_{110} + C_{101} - D_{100} + E_{011} + F_{010} + G_{001} + H_{000}$	$-\sigma_x \otimes F_1 \otimes F_1 - I\sigma_y \otimes F_2 \otimes F_1 + \sigma_x \otimes \sigma_x \otimes F_2$

State Obtained by Alice	Short-Hand Form of Transformation
$-A_{111} + B_{110} - C_{101} - D_{100} + E_{011} + F_{010} + G_{001} + H_{000}$	$I\sigma_y \otimes F_1 \otimes F_1 - \sigma_x \otimes F_2 \otimes F_1 + \sigma_x \otimes \sigma_x \otimes F_2$
$A_{111} - B_{110} - C_{101} - D_{100} + E_{011} + F_{010} + G_{001} + H_{000}$	$-I\sigma_y \otimes F_1 \otimes F_1 - \sigma_x \otimes F_2 \otimes F_1 + \sigma_x \otimes \sigma_x \otimes F_2$
$-A_{111} - B_{110} - C_{101} - D_{100} + E_{011} + F_{010} + G_{001} + H_{000}$	$-\sigma_x \otimes \sigma_x \otimes F_1 + \sigma_x \otimes \sigma_x \otimes F_2$

Author details

Mrittunjoy Guha Majumdar
Indian Institute of Science, Bangalore, India

*Address all correspondence to: mrittunjoyguhamajumdar@hotmail.com

References

[1] C. H. Bennett, G. Brassard, C. Crépeau, R. Jozsa, A. Peres, and W. K. Wootters, Teleporting an unknown quantum state via dual classical and einstein-podolskyrosen channels, Physical review letters **70**, 1895 (1993).

[2] W. Tian-Yin and W. Qiao-Yan, Controlled quantum teleportation with bell states, Chinese Physics B **20**, 040307 (2011).

[3] X. Tan, X. Li, and P. Yang, Perfect quantum teleportation via bell states, Computers, Materials & Continua **57**, 495 (2018).

[4] J. Dong and J. Teng, Controlled teleportation of an arbitrary n-qudit state using nonmaximally entangled ghz states, The European Physical Journal D **49**, 129 (2008).

[5] S. Hassanpour and M. Houshmand, Bidirectional teleportation of a pure epr state by using ghz states, Quantum Information Processing **15**, 905 (2016).

[6] X. Gao, Z. Zhang, Y. Gong, B. Sheng, and X. Yu, Teleportation of entanglement using a three-particle entangled w state, JOSA B **34**, 142 (2017).

[7] B.-S. Shi and A. Tomita, Teleportation of an unknown state by w state, Physics Letters A **296**, 161 (2002).

[8] P. Espoukeh and P. Pedram, Quantum teleportation through noisy channels with multi-qubit ghz states, Quantum information processing **13**, 1789 (2014).

[9] R.-G. Zhou, C. Qian, and H. Ian, Cyclic and bidirectional quantum teleportation via pseudo multi-qubit states, IEEE Access **7**, 42445 (2019).

[10] A. Kumar, S. Adhikari, S. Banerjee, and S. Roy, Optimal quantum communication using multiparticle partially entangled states, Physical Review A **87**, 022307 (2013).

[11] X.-Z. Zhou, X.-T. Yu, and Z.-C. Zhang, Multi-hop teleportation of an unknown qubit state based on w states, International Journal of Theoretical Physics **57**, 981 (2018).

[12] D. Joy and M. Sabir, Efficient schemes for the quantum teleportation of a sub-class of tripartite entangled states, Quantum Information Processing **17**, 170 (2018).

[13] Q. Quan, M.-J. Zhao, S.-M. Fei, H. Fan, W.-L. Yang, and G.-L. Long, Two-copy quantum teleportation, Scientific reports **8**, 1 (2018).

[14] S. Rajiuddin, A. Baishya, B. K. Behera, and P. K. Panigrahi, Experimental realization of quantum teleportation of an arbitrary two-qubit state using a four-qubit cluster state, Quantum Information Processing **19**, 87 (2020).

[15] Y.-H. Luo, H.-S. Zhong, M. Erhard, X.-L. Wang, L.-C. Peng, M. Krenn, X. Jiang, L. Li, N.-L. Liu, C. Y. Lu, et al., Quantum teleportation in high dimensions, Physical review letters **123**, 070505 (2019).

[16] K. Hofmann, A. Semenov, W. Vogel, and M. Bohmann, Quantum teleportation through atmospheric channels, Physica Scripta **94**, 125104 (2019).

[17] V. Sharma, C. Shukla, S. Banerjee, and A. Pathak, Controlled bidirectional remote state preparation in noisy environment: a generalized view, Quantum Information Processing **14**, 3441 (2015).

[18] Y.-J. Duan, X.-W. Zha, X.-M. Sun, and J.-F. Xia, Bidirectional quantum controlled teleportation via a maximally

seven-qubit entangled state, International Journal of Theoretical Physics **53**, 2697 (2014).

[19] F.-G. Deng, C.-Y. Li, Y.-S. Li, H.-Y. Zhou, and Y.Wang, Symmetric multiparty-controlled teleportation of an arbitrary two-particle entanglement, Physical Review A **72**, 022338 (2005).

[20] F. Yan and D. Wang, Probabilistic and controlled teleportation of unknown quantum states, Physics Letters A **316**, 297 (2003).

[21] Z.-j. Zhang and C.-Y. Cheung, Shared quantum remote control: quantum operation sharing, Journal of Physics B: Atomic, Molecular and Optical Physics **44**, 165508 (2011).

[22] Q. Ji, Y. Liu, X. Yin, X. Liu, and Z. Zhang, Quantum operation sharing with symmetric and asymmetric w states, Quantum information processing **12**, 2453 (2013).

[23] M. Hillery, V. Bužek, and A. Berthiaume, Quantum secret sharing, Physical Review A **59**, 1829 (1999).

[24] L. Xiao, G. L. Long, F.-G. Deng, and J.-W. Pan, Efficient multiparty quantum-secret-sharing schemes, Physical Review A **69**, 052307 (2004).

[25] D. Gottesman, Theory of quantum secret sharing, Physical Review A **61**, 042311 (2000).

[26] J. Heo, C.-H. Hong, M.-S. Kang, H. Yang, H.-J. Yang, J.-P. Hong, and S.-G. Choi, Implementation of controlled quantum teleportation with an arbitrator for secure quantum channels via quantum dots inside optical cavities, Scientific reports **7**, 1 (2017).

[27] T. Zheng, Y. Chang, and S.-B. Zhang, Arbitrated quantum signature scheme with quantum teleportation by using two three-qubit ghz states, Quantum Information Processing **19**, 1 (2020).

[28] M. Walter, D. Gross, and J. Eisert, Multipartite entanglement, Quantum Information: From Foundations to Quantum Technology Applications, **293** (2016).

[29] T. Dash, R. Sk, and P. K. Panigrahi, Deterministic joint remote state preparation of arbitrary two-qubit state through noisy cluster-ghz channel, Optics Communications **464**, 125518 (2020).

[30] Z.-Z. Zou, X.-T. Yu, Y.-X. Gong, and Z.-C. Zhang, Multihop teleportation of two-qubit state via the composite ghz–bell channel, Physics Letters A **381**, 76 (2017).

[31] S. Shuai, N. Chen, and B. Yan, Bidirectional quantum communication through the composite ghz-ghz channel, Applied Sciences **10**, 5500 (2020).

[32] Y.-y. Nie, Y.-h. Li, X.-p. Wang, and M.-h. Sang, Controlled dense coding using a five-atom cluster state in cavity qed, Quantum information processing **12**, 1851 (2013).

[33] X.-J. Yi and J.-M. Wang, Spin squeezing of superposition of multi-qubit ghz state and w state, International Journal of Theoretical Physics **50**, 2520 (2011).

[34] L. Chen and Y.-X. Chen, Classification of ghz-type, wtype, and ghz-w-type multiqubit entanglement, Physical Review A **74**, 062310 (2006).

[35] Z.-X. Man, Y.-J. Xia, and N. B. An, Genuine multiqubit entanglement and controlled teleportation, Physical Review A **75**, 052306 (2007).

[36] C.-P. Yang, Q.-P. Su, Y. Zhang, and F. Nori, Implementing a multi-target-qubit controlled-not gate with logical qubits outside a decoherence-free

subspace and its application in creating quantum entangled states, Physical Review A **101**, 032329 (2020).

[37] P. Contreras-Tejada, C. Palazuelos, and J. I. De Vicente, Resource theory of entanglement with a unique multipartite maximally entangled state, Physical Review Letters **122**, 120503 (2019).

[38] T. Theurer, N. Killoran, D. Egloff, and M. B. Plenio, Resource theory of superposition, Physical review letters **119**, 230401 (2017).

[39] S. Bäuml, S. Das, X. Wang, and M. M. Wilde, Resource theory of entanglement for bipartite quantum channels, arXiv preprint arXiv: 1907.04181 (2019).

[40] F. Shahandeh, The resource theory of entanglement, in Quantum Correlations (Springer, 2019) pp. 61–109.

[41] D. A. Lidar, I. L. Chuang, and K. B. Whaley, Decoherence-free subspaces for quantum computation, Physical Review Letters **81**, 2594 (1998).

[42] D. Bacon, J. Kempe, D. A. Lidar, and K. B. Whaley, Universal fault-tolerant quantum computation on decoherence-free subspaces, Physical Review Letters **85**, 1758 (2000).

[43] P. G. Kwiat, A. J. Berglund, J. B. Altepeter, and A. G. White, Experimental verification of decoherence-free subspaces, Science **290**, 498 (2000).

[44] D. A. Lidar and K. B. Whaley, Decoherence-free subspaces and subsystems, in Irreversible quantum dynamics (Springer, 2003) pp. 83–120.

[45] J. Roffe, Quantum error correction: an introductory guide, Contemporary Physics **60**, 226 (2019).

[46] R. Raussendorf and J. Harrington, Fault-tolerant quantum computation with high threshold in two dimensions, Physical review letters **98**, 190504 (2007).

[47] R. Cleve, Quantum stabilizer codes and classical linear codes, Physical Review A **55**, 4054 (1997).

[48] D. Gottesman, Stabilizer codes and quantum error correction, arXiv preprint quant-ph/9705052 (1997).

[49] R. Raussendorf and H. J. Briegel, A one-way quantum computer, Physical Review Letters **86**, 5188 (2001).

[50] S. Bravyi and R. Raussendorf, Measurement-based quantum computation with the toric code states, Physical Review A **76**, 022304 (2007).

[51] R. Raussendorf, D. E. Browne, and H. J. Briegel, Measurement-based quantum computation on cluster states, Physical review A **68**, 022312 (2003).

[52] H. J. Briegel and R. Raussendorf, Persistent entanglement in arrays of interacting particles, Physical Review Letters **86**, 910 (2001).

[53] T. J. Osborne and F. Verstraete, General monogamy inequality for bipartite qubit entanglement, Physical review letters **96**, 220503 (2006).

[54] X.-N. Zhu and S.-M. Fei, Entanglement monogamy relations of qubit systems, Physical Review A **90**, 024304 (2014).

[55] Y.-K. Bai, M.-Y. Ye, and Z. Wang, Entanglement monogamy and entanglement evolution in multipartite systems, Physical Review A **80**, 044301 (2009).

[56] M. Koashi and A. Winter, Monogamy of quantum entanglement and other correlations, Physical Review A **69**, 022309 (2004).

[57] Z.-X. Jin and S.-M. Fei, Tighter entanglement monogamy relations of qubit systems, Quantum Information Processing **16**, 77 (2017).

[58] C. Di Franco, M. Paternostro, and M. Kim, Nested entangled states for distributed quantum channels, Physical Review A 77, 020303 (2008).

[59] F. Fröwis and W. Dür, Stable macroscopic quantum superpositions, Physical review letters **106**, 110402 (2011).

[60] M. Christandl, N. Datta, A. Ekert, and A. J. Landahl, Perfect state transfer in quantum spin networks, Physical review letters **92**, 187902 (2004).

[61] M. Christandl, N. Datta, T. C. Dorlas, A. Ekert, A. Kay, and A. J. Landahl, Perfect transfer of arbitrary states in quantum spin networks, Physical Review A **71**, 032312 (2005).

[62] D. Saha and P. K. Panigrahi, N-qubit quantum teleportation, information splitting and superdense coding through the composite ghz–bell channel, Quantum Information Processing **11**, 615 (2012).

[63] S. Zippilli, J. Li, and D. Vitali, Steady-state nested entanglement structures in harmonic chains with single-site squeezing manipulation, Physical Review A **92**, 032319 (2015).

[64] A. G. Fowler, A. M. Stephens, and P. Groszkowski, Highthreshold universal quantum computation on the surface code, Physical Review A **80**, 052312 (2009).

[65] D. S. Wang, A. G. Fowler, and L. C. Hollenberg, Surface code quantum computing with error rates over 1%, Physical Review A **83**, 020302 (2011).

[66] H. Bombín and M. A. Martin-Delgado, Quantum measurements and gates by code deformation, Journal of Physics A: Mathematical and Theoretical **42**, 095302 (2009).

[67] H. Bombin, Clifford gates by code deformation, New Journal of Physics **13**, 043005 (2011).

[68] E. Dennis, A. Kitaev, A. Landahl, and J. Preskill, Topological quantum memory, Journal of Mathematical Physics **43**, 4452 (2002).

[69] M. Mhalla and S. Perdrix, Graph states, pivot minor, and universality of (x, z)-measurements, arXiv preprint arXiv:1202.6551 (2012).

[70] R. Beals, S. Brierley, O. Gray, A. W. Harrow, S. Kutin, N. Linden, D. Shepherd, and M. Stather, Efficient distributed quantum computing, Proceedings of the Royal Society A: Mathematical, Physical and Engineering Sciences **469**, 20120686 (2013).

[71] D. Bruß and N. Lütkenhaus, Quantum key distribution: from principles to practicalities, Applicable Algebra in Engineering, Communication and Computing **10**, 383 (2000).

[72] P. Komar, E. M. Kessler, M. Bishof, L. Jiang, A. S. Sørensen, J. Ye, and M. D. Lukin, A quantum network of clocks, Nature Physics **10**, 582 (2014).

[73] Z. Eldredge, M. Foss-Feig, J. A. Gross, S. L. Rolston, and A. V. Gorshkov, Optimal and secure measurement protocols for quantum sensor networks, Physical Review A **97**, 042337 (2018).

[74] T. Proctor, P. Knott, and J. Dunningham, Networked quantum sensing, arXiv preprint arXiv: 1702.04271 (2017).

[75] W. Ge, K. Jacobs, Z. Eldredge, A. V. Gorshkov, and M. Foss-Feig, Distributed quantum metrology with linear networks and separable inputs,

Physical review letters **121**, 043604 (2018).

[76] V. Kuzmin, D. Vasilyev, N. Sangouard, W. Dür, and C. Muschik, Scalable repeater architectures for multiparty states, npj Quantum Information **5**, 1 (2019).

[77] M. G. Majumdar and S. S. Garani, Quantum network recovery from multinode failure using network encoding with ghz-states on higher-order buttery networks, arXiv preprint arXiv:2101.01541 (2021).

[78] X.-W. Zha, H.-Y. Song, J.-X. Qi, D. Wang, and Q. Lan, A genuine maximally seven-qubit entangled state, arXiv preprint arXiv:1110.5011 (2011).

[79] I. D. Brown, S. Stepney, A. Sudbery, and S. L. Braunstein, Searching for highly entangled multi-qubit states, Journal of Physics A: Mathematical and General **38**, 1119 (2005).

[80] A. Borras, A. Plastino, J. Batle, C. Zander, M. Casas, and A. Plastino, Multiqubit systems: highly entangled states and entanglement distribution, Journal of Physics A: Mathematical and Theoretical **40**, 13407 (2007).

[81] Y. Chen, Bidirectional controlled quantum teleportation by using five-qubit entangled state, International Journal of Theoretical Physics **53**, 1454 (2014).

[82] X. Tan, X. Zhang, and T. Song, Deterministic quantum teleportation of a particular six-qubit state using six-qubit cluster state, International Journal of Theoretical Physics **55**, 155 (2016).

[83] R.-G. Zhou, R. Xu, and H. Lan, Bidirectional quantum teleportation by using six-qubit cluster state, IEEE Access 7, 44269 (2019).

[84] M.-h. Sang, Bidirectional quantum controlled teleportation by using a seven-qubit entangled state, International Journal of Theoretical Physics **55**, 380 (2016).

[85] Y.-J. Duan, X.-W. Zha, X.-M. Sun, and J.-F. Xia, Bidirectional quantum controlled teleportation via a maximally seven-qubit entangled state, International Journal of Theoretical Physics **53**, 2697 (2014).

[86] W.-q. Hong, Asymmetric bidirectional controlled teleportation by using a seven-qubit entangled state, International Journal of Theoretical Physics **55**, 384 (2016).

[87] M. S. S. Zadeh, M. Houshmand, and H. Aghababa, Bidirectional teleportation of a two-qubit state by using eight-qubit entangled state as a quantum channel, International Journal of Theoretical Physics **56**, 2101 (2017).

Chapter 4

Some Foundational Issues in Quantum Information Science

Amitabha Gupta

Abstract

This chapter has three Parts. Part 1 attempts to analyze the concept "information" (in some selected contexts where it has been used) in order to understand the consequences of representing and processing *information, quantum mechanically*. There are at least three views on 'Information' viz., 'Semantic Naturalism', 'the Quantum Bayesian Approach' and 'Information is Physical' approach. They are then critically examined and at last one is given preference. Part 2 of the chapter then goes on to discuss the manner in which the study and quantification of "*Qubit*" *(Quantum bit)*, Superposition and Entanglement, comprise the three pillars of Quantum Information Science and enable the discipline to develop the theory behind applications of quantum physics to the transmission and processing of information. In Part 3 we take up the issue that although it might appear bewildering, the physical approach to Quantum Information Science is equally proficient in dealing with its impact on the questions of "consciousness," "freewill" and biological questions in the area known as **"bioinformatics."**

Keywords: Meaning and Types of Information, The role of 'Qubit', 'Superposition' and 'entanglement' in Quantum Information Science and their applications, microtubules, Orch OR Model, biological theories of mind

1. Introduction

Quantum Information Science addresses the question as to how the fundamental laws of quantum physics can be exploited in order to explain in what way information is acquired, transmitted and processed, by drawing insights from various subfields of physical sciences, computer science, mathematics, and engineering. Quantum Information Science also combines fundamental research with practical applications.

The history of quantum information began at the turn of the 20th century when classical physics was revolutionized into quantum physics. The field of quantum information bloomed two decades ago when scientists realized that quantum physics could be used to transmit and process information in more efficient and secure ways. The development of quantum algorithm and communication protocols as well as the possibilities of implementing them with different systems, has established the field of quantum information science as one of the most promising fields for the 21st century.

The emergence of Information theory as studies of the transmission, processing, extraction, and utilization of information received immediate worldwide attention in the late forties. It was made possible by the publication of Claude E. Shannon's classic paper "A Mathematical Theory of Communication" [1].

Shannon for the first time introduced the qualitative and quantitative models of communication as statistical processes underlying information theory. Thus, Information theory often concerns itself with measures of information distributions and their application.

There is an urgent need to examine the foundational principles of quantum information and quantum physics in order to understand how we can dramatically improve their applications. The relevant utility of quantum computers has led to the possibility of simulating the complex quantum systems that appear in fields, such as condensed matter physics, high energy physics or chemistry. To do this, it is often necessary to build a scalable quantum computer (often called quantum simulators) and not necessarily an analog one. Quantum information science also has strong connections with quantum sensing and metrology, quantum simulation, quantum networks, and quantum dynamics. Issues in Quantum Information Science also found applications in areas, including statistical inference, cryptography, neurobiology, perception, linguistics, bioinformatics, quantum computing, information retrieval, plagiarism detection, pattern recognition, anomaly detection, biology and many other areas.

2. "Information" in quantum information science: What does the word mean?

Understanding the concept "information" is of importance to all the information disciplines. Perhaps for that reason, it is a term that has been defined in countless ways, over many areas in Quantum Information Science. It would be fair to say that there is no widely agreed-upon definition or theoretical conception of the term.

According to Luciano Floridi [2] the word "information" commonly refers to at least four kinds of mutually compatible phenomena:

- Information in something (e.g. a pattern or a constraint).
- Information about something (e.g. a train timetable)
- Information as something (e.g. DNA, or fingerprints)
- Information for something (e.g. algorithms or instructions)

These four phenomena commonly referred to by the word "information" are used so metaphorically or in such an abstract way that the meaning of the word "information" looks quite unclear. In spite of this lack of clarity in the meaning of information, Floridi was primarily concerned with efforts in rethinking and "reengineering" our societal concerns for "information" in the digital age.

Floridi's main interest in Quantum Information Science relates to the problems raised by analysis of "information" in two different areas:

a. Information Theory that takes into account (i) technical problems concerning the quantification of information first dealt with by Shannon's theory and (ii) the problems concerning the impact and effectiveness of information on human behavior and

b. semantic problems (semantics is concerned with how we derive meaning) relating to meaning of life, truth and being Human in a "Digital reality" or "Hyperconnected Era" as the development and widespread use of information communication technologies (ICTs) are having a radical impact on the human condition.

In view of the lack of agreement about the definition of the term "information," as shown above, the main objective in Part 1, will be to lay out some of the major theoretical constructions giving rise to the classes of definitions of the term "information" that are currently or recently in use. What binds these theoretical constructions together is the claim that the technical notion of "information" can be specified only by using a purely mathematical and physical vocabulary. In what follows, we will discuss a few theoretical constructions.

2.1 The semantic naturalism

"Semantics" involves the **understanding of the relationship between** words, **texts** signals, sentence structure on the one hand **and language-independent, reference, truth and meaning on the other. There are** three types of semantics: **Formal, Lexical and Conceptual Semantics. "Semantic Naturalism" is essentially the view that it is possible to have physical understanding of "meaning."**

There is in philosophy a tradition occupied by those who hope or expect to achieve the reduction of semantics and related concepts to respectable **physical ones**. Daniel C. Dennett [3] introduces the concept **"intentional stance"** in which the behavior of a system is understood in terms of its goals, beliefs and a principle of rationality: it does what it believes will achieve its goals. Much of what we know, the behavior of many systems is **intentional** (i.e. pointing beyond itself or the capacity of the mind to refer to an existent or nonexistent object).

John Searle [4, 5] argues that mere calculation does not, of itself, evoke conscious mental attributes, such as understanding or intentionality, e.g. the results of mathematical insight, do not seem to be obtained algorithmically. In his famous "Chinese room argument," Searle claims to demonstrate that computers mimic someone who understands Chinese even though it does not know a word. Computers process symbols in ways that simulate human thinking, but they are actually mindless, as they do not have any subjective, conscious experience.

Fred Dretske in his **Knowledge and the Flow of Information** Dretske [6], enunciates the idea of **semantic naturalism. "Naturalism"** or **"naturalistic characterization"** is a tendency attempting to explain everything in terms of **nature** or a tendency consisting essentially of looking upon nature as the original and fundamental source of all explanation, in this case "information".

This is also Dretske's first major defense of an **informational theory of content**. This book was instrumental in bringing informational approaches to the attention of mainstream philosophers of mind. Dretske's distinctive claim in his **communication-theoretic notion of information** is that a satisfactory semantic concept of information is indeed to be found and may be articulated with a simple extension of **Shannon's theory** of **"information."** (discussed below).

The main thrust of Dretske approach, by taking advantage of the new physics, is to show how the idea of a non-algorithmic conscious brain, is capable of filling the so-called 'gap' between **physical** and **semantic (or intentional) concepts**. Extending this idea with the help of the present state of **physical understanding** of semantic or intentional concepts in *physical terms*, Fred Dretske initiated a new tradition of what we might call the **semantic naturalism**. Dretske holds the view that to **possess "information"** is to have certain capacity or ability, while for something to **contain "information"** is for it to be in a certain state or to possess certain occurrent categorical properties.

The earliest systematic attempts to understand **semantic content** or **intentional content** (i.e. the **context** in which an utterance is made or referring to **what a sentence or utterance is** and its **use**) in terms of **"information"** was carried out by Dretske.

Dretske articulates his **notion of information** and defines it in following way: a state type T carries information of type p if there is a nomological or counterfactual regularity (perhaps a ceteris paribus law) to the effect that if a T occurs, p obtains. So, for example, the height of mercury in a thermometer carries **information** about the ambient temperature. Dretske's idea is to construct the content of beliefs out of the information that they carry under certain circumstances.

Thus, a signal correlated with p will fail to carry the information that p if the correlation is merely accidental or statistical: my thermometer carries information about the temperature of my room and not somebody else's room, even if the two rooms have the same temperature. It is because the state of my thermometer supports counterfactuals about the temperature of my room but not about the temperature of somebody else's room. Hence, it is a true generalization that if the temperature of my room were different, the state of my thermometer would be different. In contrast, it is not generally true that if the temperature of somebody else's room were different, the state of my thermometer would be different.

According to Dretske the engineering aspects of mechanical communication systems are relevant and he goes on to demonstrate precisely what their relevance is. Dretske's proposal is linking the information theory to the amount of information that an individual event carries about another event or state of affairs. He argues that if a signal is to carry the information that q it must, among other things, carry *as much information* as is generated by the obtaining of the fact that q. He says:

> *How, for example, do we calculate the amount of information generated by Edith's playing tennis?... [O]ne needs to know: (1) the alternative possibilities... (2) the associated probabilities ... (3) the conditional probabilities ... Obviously, in most ordinary communication settings one knows none of this. It is not even very clear whether one could know it. What, after all, are the alternative possibilities to Edith's playing tennis? Presumably there are some things that are possible (e.g., Edith going to the hairdresser instead of playing tennis) and some things that are not possible (e.g., Edith turning into a tennis ball), but how does one begin to catalog these possibilities? If Edith might be jogging, shall we count this as one alternative possibility? Or shall we count it as more than one, since she could be jogging almost anywhere, at a variety of different speeds, in almost any direction? ([6], p. 53)*

There might be problems in specifying absolute amounts of information; but it is comparative amounts of information with which Dretske is concerned, in particular, with whether a signal carries as much information as is generated by the occurrence of a specified event, whatever the absolute values might be.

2.2 Criticisms

In our system of communication and information thus far, there is an apparent failure to provide a **satisfactory naturalized account of meaning or semantics**. The important reason for this is that language, being a **rule governed** activity, has an essential **normative component** that cannot be captured by any **naturalistic explanation**. The impetus behind this line of thought derives from Wittgenstein's reflections on meaning and rule-following ([7], p. 53).

Secondly, we learn from the circumstances certain beliefs in which the information these beliefs carry is not the belief's content. Take for example a child who learns to token a belief with a content about tigers by seeing pictures of tigers. In such cases her belief states carry information about pictures, in spite of the fact that their content is about tigers. Dretske's account will end up assigning the wrong truth conditional contents to these beliefs.

Thirdly, according to Dretske a teleological characterization of the state tokens, although the relevant information fixes the content of the beliefs. However, Dretske's main idea is that the informational content fixes the belief's semantic content in these instances of the belief state and they are reinforced by the relevant behavior which produces them. Although this is a naturalistic characterization of this class of beliefs, it is debatable whether it assigns appropriate contents. One may easily come up with situations in which a false token of a belief produces behavior.

Finally, it is believed that informational theories are the most promising proposals for reconciling naturalism with intentional realism. However, it remains to be shown that there is an informational theory of content that satisfies the constraint, viz. `ps cause Ss' is a law (where S has property p as its content). Of course, this does not mean that no informational theory can succeed. However, it does mean that, so far, appeals to information have not resolved the problem of naturalizing content.

3. The quantum theory as usually presented in terms of Bayesian approach

> *"To suppose that, whenever we use a singular substantive [e.g. "information], we are, or ought to be, using it to refer to something, is an ancient, but no longer a respectable, error."* [8],

Taking Strawson's stricture of construing 'information' as a singular substantive, disembodied abstract entity as "an ancient, but no longer a respectable error," we need to look at **the probability approach** as a possible alternative.

The fallacy of construing information as a disembodied abstract entity can be avoided by taking *information as a range of possible results with varying probabilities*. Abstractly, information can be thought of as the resolution of uncertainty. While probability theory allows us to make uncertain statements and reason in the presence of uncertainty, information allows us to quantify the amount of uncertainty in a probability distribution, Let take an example, suppose we wish to compute the probability whether a poker player will win a game provided she possesses certain set of cards, exactly the same probability formulas would be used in order to compute the probability that a patient has a specific disease when we observe that she has certain symptoms. The reason for this is as follows: Probability theory provides as a set of formal rules for determining the likelihood of a proposition being true given the likelihood of other propositions.

It was in the second half of the 18th century, there was no branch of mathematical sciences called "Probability Theory". It was known simply by a rather odd-sounding "*Doctrine of Chances*". An article called, "*An Essay towards solving a Problem in the Doctrine of Chances*", authored by Thomas Bayes [9], was read to Royal Society and published in the *Philosophical Transactions of the Royal Society of London,* in 1763.

In this essay, Bayes described a simple theorem concerning joint probability which gives rise to the calculation of inverse probability. This is called Bayes' Theorem. It shows that there is a link between *Bayesian inference* and *information theory* that is useful for model selection, assessment of information entropy and experimental design.

$$\overset{\text{Posterior probabity}}{p(A|B)} = \frac{\overset{\text{Likelihood}}{p(B|A)} \quad \overset{\text{Prior probablity}}{p(A)}}{p(B)} \tag{1}$$

What is conveyed by this formula is that we update our belief, (i.e. *prior probability*), after observing the data/evidence (or the *likelihood*) of the belief and assign the updated degree of belief the term *posterior probability*. Our starting point could be a belief, however each data point will either strengthen or weaken that belief and this is how we update our belief or hypothesis all the time.

3.1 Objective certainty in finite probability spaces

In 1935, Einstein, Podolsky, and Rosen made the following sufficient condition for reality. Einstein, Podolsky and Rosen maintain that

> *"If, without in any way disturbing a system, we can predict with certainty (i.e., with probability equal to unity) the value of a physical quantity, then there exists an element of physical reality corresponding to this physical quantity. [In this sense], can quantum-mechanical description of physical reality be considered complete?" [10],*

Rudolf Carnap and Yehoshua Bar-Hillel, also in Carnap Rudolf and Bar-Hillel Yehoshua [11] '*An Outline of a Theory of Semantic Information*', take a probabilistic approach that capitalizes on the notion of the uncertainty of a piece of information in a given probability space.

3.2 The essential claim of quantum Bayesian approach

Quantum theory (as usually presented with the Born Rule, in its simplest form), states that the probability density of finding a particle at a given point, when measured, is proportional to the square of the magnitude of the particle's wave function at that point. It provides an algorithm for generating probabilities for alternative outcomes of a measurement of one or more observables on a quantum system. Traditionally they are regarded as objective. On the other hand, a subjective Bayesian or personalist view of quantum probabilities regard quantum state assignments as subjective.

3.3 Critical remarks

At the turn of the 21st century Quantum Bayesianism emerged as a result of the collaborative work among Caves et al. [12].

First, the word "Bayesian" does not carry a commitment to denying objective probability and a "Quantum Bayesian" insists that probability has no physical existence even in a quantum world. The probability ascriptions arise from a particular state that are understood in a purely Bayesian manner. Caves, Fuchs, and Schack refute Einstein, Podolsky, and Rosen's argument that quantum description is incomplete by giving up all objective physical probabilities. They would rather identify probability 1 with an **agent's subjective certainty**.

Secondly, the quantum state ascribed to an individual system is understood to represent a compact summary of **an agent's degrees of belief** about what the results of measurement interventions on a system are. Thus, an agent's degree of belief in terms of Quantum Bayesian approach is quite **subjective** and hence it would be characterized by a **non-realist view** of the quantum state.

4. 'Information is Physical' Approach: An alternative

The fact that 'information is physical' means and that the laws of Quantum Mechanics can be used to process and transmit it in ways that are not possible with classical systems.

Thus, *Classical Information Theory* is the mathematical theory of information that involves processing tasks such as storage and transmission of information, whereas *Quantum Information Theory* is the study of how such tasks can be accomplished using quantum mechanical systems.

4.1 Foundational issues

Quantum Physics, ever since it was advanced in the 1920s, has led to countless discussions about its meaning and about how to interpret the theory correctly. These discussions relate to the issues like the Einstein-Podolsky-Rosen paradox, quantum non-locality and the role of measurement in quantum physics and several others. For example, in stating their paradox on the basis of a certain restricted set of correlations for a pair of systems in a particular entangled state (explained below), Einstein et al. [10], claimed that the phenomenon of entanglement conflicts with certain basic realist principles of separability and locality that all physical theories should respect. Otherwise we have to regard quantum states as 'incomplete' descriptions of reality.

Challenging Einstein in 1927 during the **fifth Solvay Conference** (from October 24 to 29), on Electrons and Photons, which championed Quantum Theory, physicist Niels Bohr argued that the mere act of indirectly observing the atomic realm changes the outcome of quantum interactions. Nevertheless, according to Bohr, quantum predictions based on probability accurately describe reality. The so-called *Copenhagen interpretation*, which is a collection of views about the meaning of quantum mechanics principally attributed to Niels Bohr and Werner Heisenberg, also emerged in 1927. Bohr presented his view on quantum mechanics for the first time and Bohr's presentation of his view on quantum mechanics came to be called the Copenhagen interpretation, in honor of Bohr's home city. It combined his own idea of particle-wave complementarity with Max Born's probability waves and Heisenberg's uncertainty principle.

Earlier around 1926, Erwin Schrödinger had already developed a mathematical formula to describe such "matter waves", which he pictured as some kind of rippling sea of smeared-out particles. But Max Born showed that Schrödinger's waves are, in effect, "waves of probability". They encode the statistical likelihood that a particle will show up at a given place and time based on the behavior of many such particles in repeated experiments. When the particle is observed, something strange appears to happen: the wave-function "collapses" to a single point, allowing us to see the particle at a particular position.

In recent years research into the very foundations of quantum mechanics has given rise to the present field, i.e. Quantum Information Science and Technology. Thus the use of quantum physics could revolutionize the way we process and communicate information. The slogan that 'Information is Physical' is often presented as the fundamental insight at the heart of quantum information theory; after all 'information' is an abstract noun referring to something physical, transmitted from one point to another and it is frequently claimed to be entailed, or at least suggested, by the theoretical and practical advances of quantum information and computation.

4.2 Claude Shannon

The concept of information and technical notions of information, is derive from the work of Claude Shannon in his **A Mathematical Theory of Communication,**

Claude E. Shannon [1], Shannon's concept of information tells us the irreducible meaning content of the message, specified in bits, which somehow possess their own intrinsic meaning. However,

> *"The fundamental problem of communication is that of reproducing at one point either exactly or approximately a message selected at another point. Frequently these messages have meaning ... These semantic aspects of communication are irrelevant to the engineering problem." ([1], p. 31)*

We must take note first that the notion of "information" in the semantic aspects of communication did not concern Shannon. His notion of "information" is often called **"mathematical information"** and it names a branch of study which deals with quantitative measures of information. For example, binary digit, or bit, can store two pieces of information, since it can represent two different states. Two bits can store four states, however: 00, 01, 10 and 11. Three bits can store eight states and so on. This can be generalized by the formula $\log_2(x)$, where x represents the number of possible symbols in the system.

Secondly, Shannon, in his mathematical theory of information, introduces the term "**entropy**." Entropy is a key measure in information theory. It quantifies the amount of uncertainty involved in the value of a random variable or the outcome of a random process. We can illustrate it by identifying the outcome of a fair coin flip with two equally likely outcomes. It therefore provides less information or **lower entropy** than specifying the outcome from a roll of a die with six equally likely outcomes.

Shannon borrowed the term "entropy" from John von Neumann. However, in Shannon's undertaking, the notion of **'Information entropy'** tells us about **the measure of the uncertainty corresponding to unpredictability of a piece of information**. Thus, it is claimed that information that is highly probable, hence, more predictable, has a lower entropy value than less distributed information, since 'less distributed information' discloses less about the world.

Finally, the important aspect of communication can be specified by bits, which signify the physical aspect of the message and yet, somehow, it carries the meaning of the message from one point to another by encoding and decoding. However, in Shannon's mathematical theory of information, the messages in question will not have meaning. For example, while we talk in a telephone what is transmitted is not what is said into the telephone, but an analogue signal. This analogue signal records the sound waves made by the speaker, which is transmitted digitally following an encoding. Thus, a communication system consists of an information source, a transmitter or encoder, (possibly noisy) a channel, and a receiver or decoder. These are the physical aspect of the message and what mainly concerns information scientists and engineers.

John Barwise and Jerry Seligman [13], identify the 'inverse relationship principle'. The inverse relationship principle says that the informativeness of a piece of information increases as its probability decreases. This position is closely linked to the notion of *information entropy*. They claim that the quantification of semantic content demonstrates a firm relationship between semantic information and the mathematical quantification of data, previously envisioned by Shannon.

4.3 Rolf Landauer

Perhaps the most vociferous proponent of the idea that 'information is physical' was the late Rolf Landauer. In the two articles by him and one related to his work, viz., Landauer Rolf [14–17], Landauer made a very important and new observation, i.e. that information is not independent of the physical laws used to store and processes it. Information is physical, or is a fundamental constituent of the universe. Landauer's point is that whenever we find information, we find it inscribed or encoded somehow in a physical medium of whatever kind.

Although modern computers rely on quantum mechanics to operate, the information itself is still encoded classically.

> *"Information is not an abstract entity but exists only through a physical representation, thus tying it to all the restrictions and possibilities of our real physical universe ... information is inevitably inscribed in a physical medium" ([17], p. 63, 64).*

Moreover, it seems that Quantum Information Theory itself provides an apt illustration of the claim that 'Information is Physical'. But why is it that this claim is being made?

Since Landauer's very first work, viz., Landauer Rolf [14], "Dissipation and heat generation in the computing process," it was argued that information has a physical nature. As Galindo and Martin-Delgado in [18], point out that information is normally printed on a physical support, it cannot be transported faster than in vacuum, and it abides by natural laws. Moreover, they maintain that the statement that information is physical does not simply mean that a computer is a physical object, but in addition that information itself is a physical entity. In turn, this implies that the laws of information are restricted or governed by the laws of physics, in particular, those of quantum physics. Thus, information is not a disembodied abstract entity; it is always tied to a physical representation.

The first important results supporting the idea that "information is physical" was Landauer's **erasure principle**. it concerns the minimum amount of energy that has to be dissipated by a computing device when erasing one bit of information, The principle also states that the erasure of information is inevitably accompanied by the generation of heat. Bennett states the Principle in the following way: Landauer's **erasure principle** claims that

> *"any logically irreversible manipulation of information, such as the erasure of a bit or the merging of two computation paths, must be accompanied by a corresponding entropy increase in non-information-bearing degrees of freedom of the information-processing apparatus or its environment" [19],*

It must be emphasized that Landauer's principle is valid both in classical and quantum physics.

Let us take a look at two ways the experts reacted to this view:

The question is: do the truly fundamental laws of nature concern, not waves and particles, but "information"?

According to one view the truly fundamental laws of Nature concern information, not waves or particles and it is taken to be the basic postulate. For example, it is known that quantum key distribution is possible but 'quantum bit' commitment is not and that nature is nonlocal (but not as nonlocal as is imposed by causality).

According to the other view: "Information is information, not matter or energy" ([20], p. 132).

This view will be supported by Shannon. For Shannon what a sender transmits to a receiver is not information but a message. While defining information Shannon is strictly concerned with the potential selections of messages or, more precisely, of the signs available in order to codify them, Shannon' theory does not come to grips with communication as transmission of meaning or with information as the meaning of a message. His theory is mainly concerned with codification and transmission of messages. It equates two terms that are apparently opposed, namely information and uncertainty. What Shannon aims to quantify is not an 'information flow,' [6], but the transmission of messages that can be continuous, discrete or mixed. This transmission is based on a medium or, more precisely, on a messenger and is understood as a formal relation between messages.

5. The view that 'Information is Physical' is the Foundation of Quantum Information Theory

Claude Shannon in a truly remarkable paper, Claude E. Shannon [1], laid down the foundations of the subject. In this paper Shannon claims that the main concern of Quantum Information Theory is as follows:

> *"The fundamental problem of communication (under quantum information theory) is that of reproducing at one point either exactly or approximately* the quantity of information *selected at another point." [1]*

The Quantum Information Theory is much richer and more complex (than its classical counterpart) and it is inherently interdisciplinary in nature, since it touches on multiple fields and brings physicists, computer scientists, and mathematicians together on common goals.

It is far from being complete but has already found application areas well beyond the processing and transmission of information. In particular, it provides a new perspective to investigate, characterize, and classify the complex behavior of large quantum systems, ranging from materials to chemical compounds, high energy problems, and even holographic principles.

Nevertheless, even if Quantum Information Theory reinforces the notion that 'Information is Physical,' based on quantum physics, the notion itself is also relevant within classical physics.

5.1 Shannon's definition of quantity of information

Shannon defined the 'quantity of information' produced by a source, for example, the quantity in a message by a formula similar to the equation that defines thermodynamic entropy in physics. In classical thermodynamics, entropy is a property of a thermodynamic system that expresses the direction or outcome of spontaneous changes in the system. According to Shannon Entropy predicts that certain processes are irreversible or impossible, despite not violating the conservation of energy.

Shannon introduced as his most basic term, viz. **informational entropy**. It is the number of binary digits required to encode a message. This might appear currently to be a simple, even obvious way to define how much information is in a message. However, in 1948, at the very origin of quantum. Information age, the digitization of information of any sort was considered to be a revolutionary step. Shannon's 1948 paper might have been the first to use the word "bit," short for binary digit.

Shannon's paper contained two theorems. The first of these is the *source coding theorem*, which gives a formula for how much a source emitting random signals can be compressed, while still permitting the original signals to be recovered with high probability.

The second theorem, *the channel coding theorem*, states that with high probability, n uses noisy channel N can communicate C – o (n) bits reliably, where C is the channel capacity.

Thus, a new approach emerges as a result of treating information as a quantum concept and to ask what new insights can be gained by encoding this information in individual quantum systems.

5.2 Generalizations and Laws in quantum information science

While we often treat **information** in abstract terms (especially in the context of computer science), it is more correct to think of **information** as being represented as different physical states and obeying the **laws** of physics.

However, what does it mean to say that information obeys the laws of physics. In Quantum Information Theory this amounts to claiming that both the transmission and processing of information are governed by **quantum laws defined in terms of "Qubits"** (and not by the classical "bits"). Since qubits behave quantumly and in terms of quantum probabilities, we can also capitalize them to explain the two most important phenomena of quantum information science, viz. "superposition" and "entanglement."

5.2.1 "Qubit"

The term for a **classical physical system** that exist in two unambiguously distinguishable states, representing 0 and 1, is often called a 'bit.' It is commonly acknowledged that the elementary quantity of information is the bit, which can take on one of two values, usually "0" and "1". If we consider any physical realization of a bit, it requires a system with two well defined states, For example in a switch *off* represents "0" and *on* represents "1". On the other hand a bit can also be represented by a certain voltage level in a logical circuit or a pit in a compact disc or a pulse of light in a glass fiber or the magnetization on a magnetic tape. For classical systems it is helpful to have the two states separated by a large energy barrier so that the value of the bit cannot change spontaneously.

On the other hand, in quantum information science, the basic variable is the **"qubit"***:* a quantum variant of the bit. In order to encode information in a two-state quantum system, it is customary to designate the two quantum states $|0\rangle$ and $|1\rangle$. The term "Qubit" seems to have been used first by Benjamin Schumacher [21], in his "Quantum coding."

Electrons possess a quantum feature called **spin**, a type of intrinsic angular momentum. In the presence of a magnetic field, the electron may exist in two possible spin states, usually referred to as 'spin up' and 'spin down'.

One of the innovative and unusual features of Quantum Information Science is the idea of "superposition" (explained below) of different states. A quantum system can be in a "superposition" of different states at the same time. Consequently, a quantum bit can be in both the $|0\rangle$ state and the $|1\rangle$ state simultaneously. This new feature has no parallel in classical information theory. Schumacher in [21], coined the word "qubit" to describe a quantum bit.

The job of the weird symbols "|" and "⟩" (the so-called the "bra-ket" notation, was introduced by Paul Dirac in [22]. It is essentially to remind us that *mathematically* we are talking about **vectors** that represent qubit states labeled 0 and 1 and *physically* they represent states of some quantum system. This helps us to distinguish them from things like the bit values 0 and 1 or the numbers 0 and 1. One way to represent this with the help of mathematics is to use two orthogonal vectors:

$$|0\rangle = \begin{bmatrix} 1 \\ 0 \end{bmatrix} |1\rangle = \begin{bmatrix} 0 \\ 1 \end{bmatrix} \tag{2}$$

Thus one of the novel features of Quantum Information Science is that a quantum system can be in a "superposition" of different states. In a sense, the quantum bit (or "qubit") can be in both the $|0\rangle$ state and the $|1\rangle$ state *at the same time*. This is one of the reasons why in 1995 Schuhmacher coined the word "qubit" to describe a quantum bit.

It is also claimed that in Quantum Computing a "qubit" carries information. The question is where is the extra information kept. The usual answer is that the extra information lies embedded in a *superposition* of *entangled states* (the two terms will be explained below). The peculiar feature of this is that any accessing of the information destroys the superposition and with it the extra information itself.

Suppose that the two vectors $|0\rangle$ and $|1\rangle$ are **orthonormal**. This means they are both orthogonal and normalized (a normalized vector is a vector in the same direction but with a norm (length) 1) and 'orthogonal' (**Figure 1**) means the vectors are at right angles):

Consider a situation where the two vectors $|0\rangle$ and $|1\rangle$ are linearly independent. This means that we cannot describe $|0\rangle$ in terms of $|1\rangle$ and vice versa. Nevertheless, it is feasible to describe all possible vectors in 2D space using both the vectors $|0\rangle$ and $|1\rangle$ and our rules of addition and multiplication by scalars,

It is maintained that the vectors $|0\rangle$ and $|1\rangle$ form a "basis" because of the fact that (i) the vectors $|0\rangle$ and $|1\rangle$ are linearly independent, and (ii) can be used to describe any vector in 2D space (**Figure 2**) using vector addition and scalar multiplication. When the vectors are both orthogonal and normalized, they construe an "orthonormal" basis.

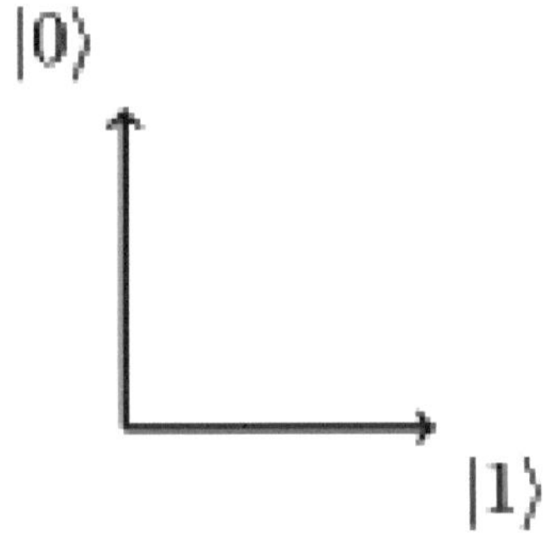

Figure 1.
'Orthogonal' vectors.

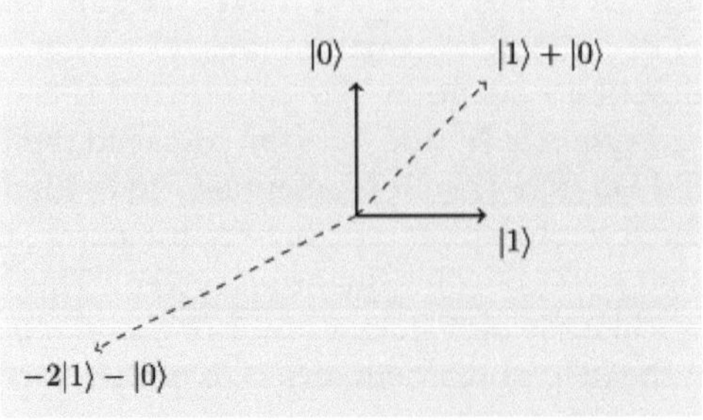

Figure 2.
Vectors in 2D space.

5.2.2 ψ function, Schrödinger equation and the dynamics of quantum mechanics

In Quantum Mechanics, the **wave function,** ψ, plays the central role and represents a variable quantity that mathematically describes the wave characteristics of a particle. At a given point of space and time the value of the wave function of a particle represents the probability of the position of the particle at the time. ψ function may be thought of as an expression for the **amplitude** of the wave of a particle. However, the spatial probability density is given by the squared modulus of the wave function, ψ^2. The Schrödinger equation is as follows:

$$ih\frac{\Psi}{t} = \frac{-h^2}{2m}\frac{{}^2\Psi}{x^2} + V(x)\Psi(x,t) \equiv \hat{\mathrm{H}}\,\Psi(\mathrm{x},\mathrm{t}) \quad (3)$$

where i is the imaginary unit, is the time-dependent wave function, is h-bar, $V(x)$ is the potential, and is the Hamiltonian operator.

The Schrödinger equation is supposed to answer the question as to how the states of a system change with time. It is in the form of a differential equation and it captures the 'dynamics' of quantum mechanics: it describes how the wave function of a physical system evolves over time.

Schrödinger's equation gives an answer to the question: what happens to the de Broglie wave associated with an electron if a force (gravitational or electromagnetic) acts on it. The equation gives the possible waves associated with the particle a number associated with any position in space at an arbitrary time (i.e. functions of position and time). The general form of this wave function is:

$$\psi(\mathrm{x},\mathrm{y},\mathrm{z};\mathrm{t}) \quad (4)$$

The essence of the Schrödinger's equation is that, given a particle and given the force system that acts (say, gravitational or electromagnetic), it yields the wave function solutions for all possible energies. Thus a particle can be described by a state vector or wave function whose evolution is provided by the Schrödinger equation. Hence, the Schrödinger equation, being a time-evolution equation, will make ψ vary with time.

5.2.3 "Qubit" and Schrödinger equation

The time-dependent **Schrödinger equation** gives the time evolution of Ψ. The entangled states are created by distributing the qubits **between** the particles so that each particle carries one **qubit**. By assuming that **a freely moving particle is the qubit carrier, it is found that both the particle position in physical space and the qubit state, change in time in accordance with the Schrödinger equation.**

5.2.4 What is quantum computing?

Basically, Quantum computing is concerned with processing information by harnessing and exploiting the amazing laws of quantum mechanics. The use of long strings of "bits" in traditional computers encode either a zero or a one. In contrast with that a quantum computer uses quantum bits, or qubits. The difference can be explained as follows: a qubit is a quantum system that encodes the zero and the one into two distinguishable quantum states. However, because qubits behave quantumly, we can capitalize on the phenomena of "superposition" and "entanglement," which is not possible in the case of using "bits," as an encoding device.

5.3 Superposition and entanglement

These two concepts might baffle us, since we do not come across the phenomena they describe in our everyday lives. Only in the event of our looking at the tiniest quantum particles, atoms, electrons, photons and the like, that we see these intriguing things, like **superposition** and **entanglement** [23].

5.3.1 Superposition

Superposition essentially subscribes to the principle that a quantum system can be in multiple states *at the same time*, that is, something can be "here" and "there," or "up" and "down" at the same time. Thus it is possible for Qubits to represent numerous possible combinations of *1* and *0* at the same time. To put qubits into superposition, researchers manipulate them using precision lasers or microwave beams. This possibility of simultaneously being in multiple states is the phenomenon of superposition. In its most basic form, this principle says that if a quantum system can be in one of two distinguishable states $|x\rangle$ and $|y\rangle$, then according to this principle it can be in any state of the form $\alpha\,|x\rangle + \beta\,|y\rangle$, where α and β are complex numbers with $|\,\alpha\,|^2 + |\,\beta\,|^2 = 1$.

5.3.1.1 Schrödinger's cat

In 1935, Erwin Schrödinger conjured up a famous **thought experiment** of putting in place a cat in a superposition of both alive and dead states. He envisioned that a cat, a small radioactive source, a Geiger counter, a hammer and a small bottle of poison were sealed in a chamber. He also imagined that if one atom of the radioactive source decays, the counter will trigger a device to release the poison. This is where Schrödinger invoked the idea of entanglement so that the state of the cat will be entangled with the state of the radioactive source. He expected that sometime after the cat will be in superposition of both alive and dead states.

It is certainly counterintuitive to think of the possibility of an organism to be in such a superposition of both alive and dead states (**Figure 3**). It also dramatically reveals the baffling consequences of quantum mechanics.

5.3.1.2 The double-slit experiment

Another well-known example of quantum superposition is the double-slit experiment in which a beam of particles passes through a double slit and forms a wave-like interference pattern on a screen on the far side.

Based on this experiment quantum interference is explained by saying that the particle is in a superposition of the two experimental paths: one passage is through

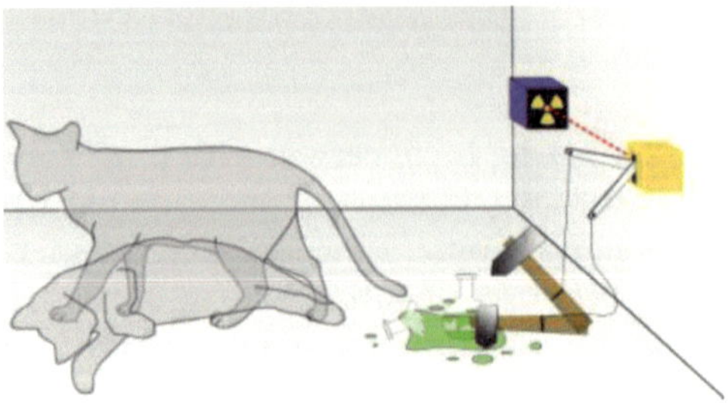

Figure 3.
Schrödinger's cat.

the upper slit and the second passage is through the lower slit. Correspondingly a quantum bit can be in a superposition of $|0\rangle$ and $|1\rangle$. The implication seems to be that each particle passes simultaneously through both slits and interferes with itself. This combination of "both paths at once" is known as a superposition state [23].

It must be noted here that the particles. After going through the two slits, will turn into two sets of waves, **Figure 4**, in accordance with quantum mechanical principles. Moreover, at some points the two sets of waves will meet crest to crest, at other points the crest will meet the trough of the other wave. Accordingly two possibilities will arise: (i) in **Figure 5**, where crest meets crest, there will be **constructive interference** and the waves will make it to the viewing screen as a bright spot, and (ii) where crest meets trough, there will be **destructive interference** that cancel each other out and a black spot will appear on the screen. One should see below bright lines of light, where the waves from the two slits *constructively interfere*, alternating with dark lines where the wave *destructively interfere*, **Figure 6**.

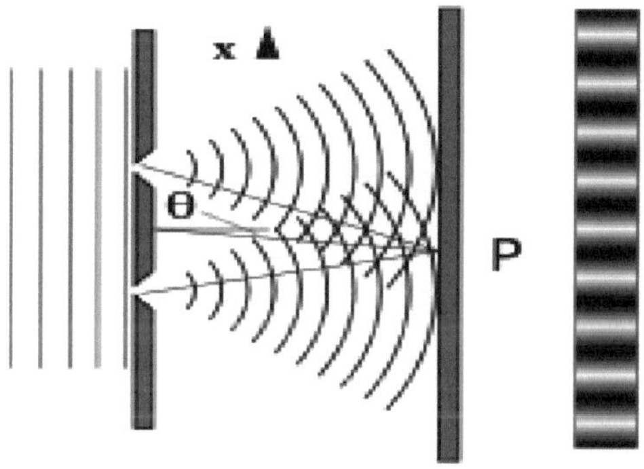

Figure 4.
Double-slit experiment.

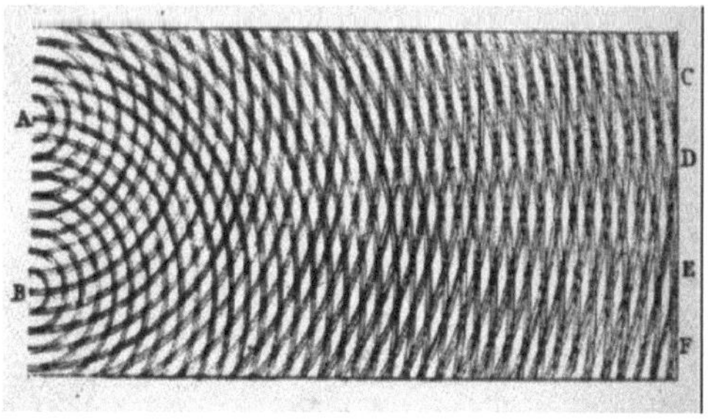

Figure 5.
Interference pattern.

Figure 6.
Constructive and destructive interference.

A particle tends to appear more often at some places (regions of constructive interference) and do not appear very often at other places (regions of destructive interference). However, the likelihood of finding the particle at a particular point can be described only probabilistically.

5.3.1.3 Superposition and quantum information science

In the two experiments explained above we have seen one of the features of a quantum system (viz, Superposition) whereby several separate quantum states can exist at the same time by superposition.

The Quantum Information Science claims that each electron will exist spin up and spin down, until it is measured. Till measurement is done it will have no chance of being in either state. Only when measured, it is **observed** to be in a specific spin state. Common experience tells us that a coin facing up is in a specific state: it is a head or a tail. Irrespective of looking at the coin, one is sure while tossing the coin must be either facing head or otherwise tail.

In **quantum experience** the situation is not as simple and more unsettling: according to quantum mechanics, **material properties of things do not exist** until **they are measured**, i.e. until one **"looks"** (measure the particular property) at the coin, as if, it has no fixed face.

5.3.2 The problem of measurement

Delving into the issue of **quantum measurement** and taking the double-slit experiment as a case in point, the "wave" of a particle, e.g., an electron, should be interpreted as relating to the *probability* of finding the particle at a specific point on the screen. We cannot detect any wave properties of a single particle in isolation. When we repeat the experiment many times, we notice that the **wave function "collapses"** and the particle is detected as a point particle. Thus, in Quantum Information Science the problem of wave function collapse is the problem of measurement of finding the *probability*.

5.3.3 Inherent uncertainty

In 1927 Heisenberg shook the physics community with his **uncertainty principle**:

$$\sigma_x \sigma_p \geq \frac{\hbar}{2} \tag{5}$$

where $\hbar$ is the reduced Planck constant, $h/2\pi$.

The formula states that the more precisely the position of some particle is determined, the less precisely its momentum can be known, and vice versa.

The uncertainty principle, the wave-particle duality, the wave collapsing into a particle when we measure it together lead to the claim that the probability of the same particle being there in several places at the same time cannot be ruled out, i.e. 'smeared out' multiple positions at a time.

The smiley face shows, **Figure** 7, the location of the particle in one peak, but then there are many such places as the multiplicity of smiley faces show.

5.3.4 Superposition and the power of a quantum computer

We have already seen (in Section 5.2.1) that whereas classical computing uses "bits" for data processing, quantum computing uses qubits. We have also

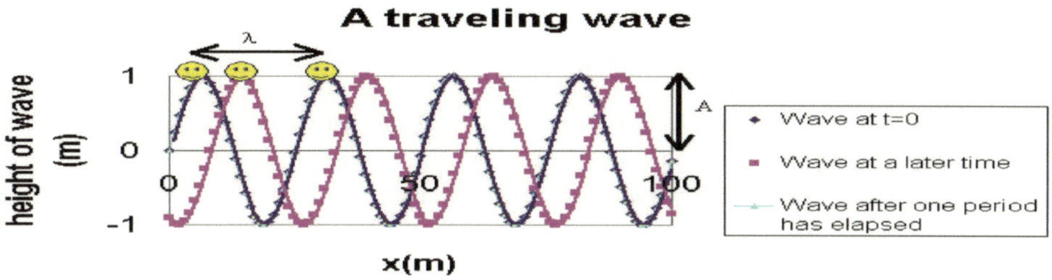

Figure 7.
A traveling wave.

scrutinized that the practical difference between a bit and a qubit is: a bit can only exist in one of two states at a time, usually represented by a 1 and a 0, whereas a qubit can exist in both states at one time.

Moreover we have observed that the phenomenon of "superposition" allows the power of a quantum computer to grow exponentially with the addition of each bit. For example, two bits in a classical computer provides four possible combinations—00, 01, 11, and 10, but only one combination at a time. Two bits in a quantum computer provides for the same four possibilities, but, because of superposition, the qubits can represent all four states at the same time, making the quantum computer four times as powerful as the classical computer. So, adding a bit to a classical computer increases its power linearly, but adding a qubit to a quantum computer increases its power exponentially: doubling power with the addition of each qubit.

5.3.5 Application of superposition in solving engineering problems

The principle of superposition is useful for solving simple practical problems, but its main use is in the **theory of circuit analysis**.

For example, in quantum science, the **superposition theorem states** that the response (**voltage** or current) in any branch of a linear circuit which has more than one independent source equals the algebraic sum of the responses caused by each independent source acting alone, while all other independent sources are turned off (made zero). Alternatively, a circuit with multiple voltage and current sources is equal to the sum of simplified circuits using just one of the sources.

5.4 Entanglement

Entanglement in quantum mechanics is considered to be an extremely strong correlation and inextricable linkage that may found between different particles of the same kind and with the same physical property. It has been observed that such linkage and intrinsic connection, subsisting between Quantum particles, is so robust, that two or more quantum particles separated albeit by great distances, may be placed at opposite ends of the universe, can still interact with each other in perfect unison. This seemingly impossible connection led Einstein to describe entanglement as "spooky action at a distance."

This intriguing phenomenon demonstrates that it is possible for scientists and researchers to generate pairs of qubits that are "entangled," which amounts also to saying that two members of a pair exist in a single quantum state. Thus, they claim that if we change the state of one of the qubits, it will bring about instantaneous change in the state of the other one in a predictable way, even if they are separated by very long distances.

The notion of entanglement was coined by Erwin Schrodinger in order to signify the peculiar properties of quantum correlations. In the *classical* world, "the whole is the sum of its parts", but the *quantum* world is very different. Schrödinger [24] says:

"the best possible knowledge of a whole *does not necessarily include the best possible knowledge of all its* parts, *even though they may be entirely separate and therefore virtually capable of being 'best possibly known,' i.e., of possessing, each of them, a representative of its own. The lack of knowledge is by no means due to the interaction being insufficiently known, at least not in the way that it could possibly be known more completely, it is due to the interaction itself.*

Attention has recently been called to the obvious but very disconcerting fact that even though we restrict the disentangling measurements to one *system, the representative obtained for the* other *system is by no means independent of the particular choice of observations which we select for that purpose and which by the way are* entirely *arbitrary. It is rather discomforting that the theory should allow a system to be steered or piloted into one or the other type of state at the experimenter's mercy in spite of his having no access to it."*

For example, consider a pair of qubits. Suppose that each one is described by a state vector: the first one by $|a\rangle$ and the second one by $|b\rangle$. One might therefore think that the most general state of the two qubits should be represented by a pair of state vectors, $|a\rangle|b\rangle$, with one for each qubit. Indeed, such a state is certainly possible, but there are other states that *cannot* be expressed in this form. The possible pair of states are also **separable** (often called **product states**). States that are not separable are said to be **entangled.** *Most* vectors are entangled and *cannot* be written as product states. This shows a peculiar feature of quantum states.

Example of entanglement when a measurement is made: a subatomic particle decays into an entangled pair of other particles. Essentially, quantum entanglement suggests that acting on a particle here, can instantly influence a particle far away. This is often described as **theoretical teleportation.** It has huge implications for quantum mechanics, quantum communication and quantum computing.

5.4.1 Entanglement and quantum information science

Quantum entangled states play a crucial role and have become the key ingredient in the field of Quantum Information Science.

It will be a fair question to ask as to why does the effect of **entanglement** matter? The answer to this is as follows: the behavior of the Quantum entangled states gives rise to seemingly paradoxical effects, viz. any measurement of a particle's properties results in an irreversible wave function collapse of that particle and changes the original **quantum** state. In the case of entangled particles such measurement will affect the **entangled system** as a whole.

Schrödinger, (unlike Einstein, the most skeptical about **entanglement** and considered it the fatal flaw in quantum theory, referring to it as "*spooky action-at-a-distance*"), was much more prepared to accept quantum theory with the concept of entanglement and along with all its predictions, no matter how weird they might be. In his paper [24], which introduced quantum entanglement, Schrödinger wrote "I would not call it *one* but rather *the* characteristic trait of quantum mechanics, the one that enforces its entire departure from classical lines of thought".

5.4.2 Application of entanglement

The interpretation of quantum states, in particular the interpretation of so-called 'entangled states' exhibit peculiar **nonlocal** (explained below) statistical correlations for widely separated quantum systems. For example, the theory underlying the field of quantum information, dealing with "entanglement," has found intriguing connections with different fields of physics, like condensed matter, quantum gravity, or string theory.

Quantum entanglement, more often is viewed as a physical resource, which enables us to communicate with perfect security, build very precise atomic clocks, and even teleport small quantum objects, dense coding and cryptography.

5.4.3 In what way entanglement enables us to communicate with perfect security

Quantum entanglement offers a new modality for communications that is different from classical communications. It has been claimed that entanglement enhances security in secret sharing.

Quantum cryptography (it is a method of storing and transmitting data quantum mechanically in a particular form so that only those for whom it is intended can read and process it) to a great extent revolves around quantum computing. The entanglement concept is one tool used in quantum computing, e.g., in the use of transmitting data via entangled Qubit, which is a unit of quantum information that is stored in a quantum system. Quantum cryptography utilizes photons and depends on the laws of physics rather than very large numbers and the deciphering of cryptographic codes.

It appears that we are perched on the edge of a quantum communication revolution that will change transmission of information, information security and how we understand privacy.

6. Nonlocality

Two central concepts of quantum mechanics are Heisenberg's uncertainty principle and nonlocality. Nonlocality plays a fundamental role in quantum information science.

Whereas the quantum entanglement, which can be traced back to the Einstein, Boris Podolsky and Nathan Rosen (EPR) paradox in 1935 (they argued that the description of physical reality provided by quantum mechanics was *incomplete*). This argument gave rise to the discussions on the foundations of quantum mechanics related to **reality** and **locality**. This plays crucial roles in quantum information processing.

Quantum Theory can predict certain patterns of correlation among spatially separated events correctly. This manifests **non-local** influences between some of these events. This is a remarkable feature of the microscopic world prescribed by quantum theory. This idea of nonlocality was described by Albert Einstein rather dismissively as "spooky action at a distance" that was mentioned above.

For example, if a pair of *electrons* is created together, one will have clockwise *spin* and the other will have anticlockwise *spin* (*spin* is a particular property of particles mentioned above). The most important point is that there are two possible states and that the entire *spin* of a quantum system must always cancel out to zero.

However, it is claimed that the two *electrons* can be considered to simultaneously have *spins* clockwise-anticlockwise and anticlockwise-clockwise respectively, under *quantum theory*, and if *superposition* is possible, If the pair are then separated by any distance (without observing and thereby **decohering** (see below) and then later checked, the second particle can be seen to instantaneously take the opposite *spin* to the first, so that the pair maintains its zero total *spin*, no matter how far apart they may be.

7. Decoherence

Quantum coherence presupposes the idea that an individual particle or object has wave functions that can be split into two separate waves. When the waves operate together in a coherent way, it is referred to as quantum coherence. Quantum decoherence means the loss of quantum coherence.

However, a quantum computer needs to operate coherently until the results are measured and read out. In implementing a quantum computer, a qubit and/or many entangled qubits must undergo unitary transformations before decoherence affects the qubit states as it no longer represents a unitary transformation. Quantum Theory gives an account of why ordinary macroscopic objects do not exhibit the interference behavior characteristic of quantum "superpositions".

8. Why do these quantum effects matter?

Simply put, they are extremely useful to the future of computing and communications technology. It is due to **superposition** and **entanglement,** a quantum computer carry out a vast number of calculations simultaneously. We know that a classical computer works with ones and zeros, however a quantum computer will have the advantage of using ones, zeros and "superpositions" of ones and zeros. Certain difficult tasks, e.g. code breaking, that have long been thought impossible (or "intractable") for classical computers will be achieved quickly and efficiently by a quantum computer.

Quantum computing is not just "faster" than classical computing, for many types of problems the quantum computer would excel, such as code breaking. The power, which is required for code breaking, is derived from quantum computing's use of "qubits" or "quantum bits."

8.1 What can a quantum computer do that a classical computer cannot?

It is easy for any computer to do factoring of large numbers or multiplying two large numbers. But calculating the factors of a very large (say, 500-digit) number, on the other hand, is considered impossible for any classical computer. In 1994, a mathematician from MIT, Peter Shor, came up with the claim that if a fully working quantum computer was available, it could factor large numbers easily.

9. Areas of application

Many experts divide technologies prompted by Quantum Information Science into three application areas: (1) Quantum Sensing and metrology, (2) Communications and (3) Computing and simulation:

9.1 Quantum sensing and metrology and quantum information science

"Quantum sensing" describes the use of a quantum system, quantum properties or quantum phenomena to perform a measurement of a physical quantity. The field of quantum sensing deals with the design and engineering of quantum sources (e.g., entangled) and quantum measurements that are able to beat the performance of any classical strategy. **Metrology**, on the other hand, is the scientific study of measurement.

Early **quantum sensors** include magnetometers based on superconducting quantum interference devices and atomic vapors, or atomic clocks. Other example of an early quantum sensor is an avalanche photodiode (ADP). ADPs have been used to detect entangled photons. Entanglement-assisted sensing, sometimes referred to as **"quantum metrology,"** or "quantum-enhanced sensing," More recently, quantum sensing has become a distinct and rapidly growing branch of research within the area of Quantum Information Science and Technology, with the most common platforms being spin qubits, trapped ions and flux qubits.

9.2 Communications, its applications and quantum information science

Quantum communications are required to increases the total computing power, especially if only processors with a few qubits are available at each network node. The most advanced application of quantum communication, and in fact of Quantum Information Processing in general, is in security. Moreover. Quantum networks provide opportunities and challenges across a range of intellectual and technical frontiers, including quantum computation and metrology.

In classical signal processing, signals traveling over fiber-optic cable about 60 miles. However, it must be retransmitted. Quantum repeaters can extend the distance the signal can be sent, but they significantly increase the complexity of the process. Communications not only must be secure, but any eavesdropping attempt will destroy the communication,

NASA developed quantum networks to support the transmission of **quantum information** for aerospace applications. This example of distribution of quantum information by NASA could potentially be utilized in secure communication. (NASA STTR 2020 Phase I SolicitationT5.04Quantum Communications). Quantum communication may provide new ways to improve communication link with security, through techniques such as quantum cryptographic key distribution. Another area of benefit is the entanglement of distributed sensor networks to provide extreme sensitivity for applications, such as astrophysics, planetary science and earth science.

9.3 Computing and simulation and quantum information science

Quantum computers have enormous potential to revolutionize many areas of our society. Quantum computing provides an exponentially larger scale than classical computing, which provides advantages for certain applications.

(a) Quantum simulation refers to the use of quantum hardware to determine the properties of a quantum system, for example, determining the properties of materials such as high-temperature superconductors, and modeling nuclear and particle physics. We have seen that harnessing quantum entanglement can solve problems more efficiently.

(b) The other approach is to simulate the behavior of quantum materials and quantum systems using controlled evolution and interaction of qubits.

10. Prophiciency of the physical approach to quantum information science in dealing with "consciousness," "freewill" and biological questions

In Part 2, we considered the physical approach to Quantum Information Science by characterizing "information" in physical terms and found it vialable. A complete physical approach to quantum information requires a robust interface among microwave photons, long-lifetime memory, and computational qubits.

It might appear to be perplexing that this physical approach to Quantum Information Science is equally proficient in dealing with "consciousness," "freewill" and biological questions in the area known as "bioinformatics."

10.1 Consciousness, quantum physics and quantum information science

One of the first processes based on which consciousness and quantum physics come together is through the Copenhagen interpretation of quantum physics. The central ideas of the Copenhagen interpretation were developed by a core group of quantum physics pioneers, centered around Niels Bohr's Copenhagen Institute in the 1920s.

According to this theory, the quantum wave function collapses due to a **conscious observer** making a measurement of a physical system. This is the interpretation of quantum physics that provoked the Schrödinger's cat thought-experiment, demonstrating some level of the absurdity of thinking that the same cat could be both alive and dead (i.e. two opposite states occurring at the same time and because of such phenomena as superposition and entanglement). Nevertheless, the claim that the quantum wave function collapses due to a **conscious observer** making a measurement of a physical system does completely match the evidence of what scientists observe at the quantum level.

This is one of the reasons why the research into consciousness forged ahead in Quantum Physics and Quantum Information Science and attempts in **understanding of human consciousness** in terms of some **physical theory**, in this case Quantum Mechanics, came to the fore.

10.1.1 Roger Penrose

Sir Roger Penrose is an English mathematical physicist, mathematician, philosopher of science and Nobel Laureate in Physics, delved deep into at least three areas in mathematical physics: gravitational radiation, the gravitational collapse of matter in the form black holes and lastly, the modeling of the universe. He touched on many subjects, such as quantum gravity, twistor theory, a new cosmology of the cosmos. However, a scientist of such repute with the vast knowledge of fundamental areas of modern physics also saw the impact and the essential role a physical theory, such as **quantum mechanics**, plays in **the understanding of human consciousness**.

The idea of using quantum physics to explain human consciousness really caught genuine interest with Roger Penrose's 1989 book, **"The Emperor's New Mind: Concerning Computers, Minds, and the Laws of Physics,"** [25]. One of Penrose's motivation to write the book was to respond to the claim of the old school of artificial intelligence researchers, who believed that the brain is capable of being modeled by "Universal Turing machine" of Alan Turing as well as the digital computers.

According to Penrose consciousness **is not** computational and **is** non-algorithmic. It is little more than a biological computer. Hence, Penrose made a distinction between his study of consciousness from any potential exploration of consciousness in artificial intelligence. In this book, Penrose argues that the brain is far more sophisticated than what the AI researchers would have us believe. The main difference is that that the brain does not operate on a strictly binary system of on and off. Instead, the human brain works with computations that are in a superposition of different quantum states at the same time. Moreover, to understand consciousness, one needs to revolutionize our understanding of the physical world.

In the initial part of the book Penrose provides a summary of Classical and Quantum Physics and argues that the physical modeling of the "real world" - from Newtonian mechanics over Einstein's relativity theory up to supersymmetry - is carried out in this Physics. However, simulation of the mind will only be possible if we understand how the missing piece of gravity radiation can be consistently included in the standard model of physics.

In the last two chapters of the book Penrose takes up his initial primary problem of modeling the human mind. To begin with Penrose gives a biophysical description of the brain and what is known about its centers and how it works. At this stage Penrose does not give a precise definition of consciousness because it is seemingly impossible. To illustrate this let us take the example of a brain, which seems to be able to register things, even when the person is **'unconscious,'** e.g. during the person undergoing an operation. We may indirectly characterize that the person's consciousness is linked to, for example, common sense judgment of truth, understanding, and artistic appraisal, whereas this is exactly opposite to automatic and algorithmic behavior. Penrose says:

> *"... neither classical nor quantum mechanics [...] can ever explain the way in which we think;" but "a plausible case can be made that there is a non-algorithmic ingredient to conscious thought processes" ([25], p. 521) and noncomputability is a feature of our conscious thinking.*

Penrose thinks that current computers will never have intelligence because of they operate under algorithmic deterministic system.

This idea is partly inspired by Penrose's experience as a mathematician and rests on **Gödel's Incompleteness theorems**. Mathematicians can know the truth of a proposition by 'insight.' The Gödel Incompleteness Theorems claim that there are **propositions that cannot be proved**. This indicates that Gödel never lost sight of the importance of human mind, which has a 'non-mechanical' and a non-computational character. "Moreover, **human beings have the ability to "see" and grasp "'truths' without proof" and have visions and intuitions in creating new knowledge and a new way of looking at things.**" [23]. In Mathematics there are, Gupta Amitabha [23, 26], at least two examples of "Mathematical Conjectures" which have been taken to be **True without any Proof**: (a) Goldbach's Conjecture (1742), which claims that every even integer greater than 2 can be expressed as the sum of two primes, and (b) Cantor's Continuuam Hypothesis (1878), which asserts that that there is no set whose cardinality is strictly between that of the integers and the real numbers.

In the last two chapters, the main concern has been as to what philosophers call the "mind–body problem". Penrose discusses the computational procedures and the noncomputational activity he assigned to the processes of consciousness, and second, he takes recourse to yet-to-be-discovered quantum-level effects to explain consciousness.

The first book of Penrose has a follow-up book, Penrose Roger [27] *Shadows of the Mind: A Search for the Missing Science of Consciousness* In this book Penrose gives examples of scientists who, by a spark of inspiration came up with a superb result, while they were not `working' on the subject following algorithmic rules. Penrose's Gödelian argument shows that humans minds are non-computable and he attempts to infer a number of claims involving consciousness and physics and ascribes consciousness to the actual physical makeup of the brain.

According to quantum mechanics, a particle can have states in which it occupies several positions at once. When we treat a system according to quantum mechanics, we have to allow for these so-called **superpositions** of alternatives. This was taken up by Penrose and he argues that these ingredients formed the basis of his follow-up book, *Shadows of the Mind*. Penrose draws from research into the molecular structures in the brain and finds suggestions of quantum-level activity that may be influencing the processing of information in the brain. Penrose found some ten thousand tubulin dimers, formed together into sheaths called **microtubules**, collections of which make up the cytoskeletons that can be thought of as the neuron's nervous system.

Penrose's collaborator, a psychologist, Stuart Hameroff also suggested that there is some biological analog of quantum computing in the brain that involves **micro-tubules** within the neurons. This idea is further developed into the so called **Orchestrated objective reduction (Orch-OR) theory**. The biological theory of mind, viz. Orch OR postulates that **consciousness originates at the quantum level inside neurons**. Instead, the conventional view is based on the idea that consciousness is a product of connections between neurons. This traditional view of consciousness relies on a mechanism of quantum process, called objective reduction.

Penrose–Hameroff theory of evolution that has made our brain the way it is and the advantage it brings to the creatures able of conscious thinking is that

> *"... neither classical nor quantum mechanics [...] can ever explain the way in which we think" but "a plausible case can be made that there is a non-algorithmic ingredient to conscious thought processes" ([27], p. 521), which is explained by the **Orch-OR theory**.*

10.2 'Free will' and quantum information science

The significance of 'free will' in quantum tests, to find a quantum perspective on "free will" leads to the issue of **conscious "free will,"** although consciousness as a causal agency or the brain mechanisms causing consciousness are unknown, and the scientific basis for consciousness, and "self," and a mechanism by which conscious agency may act in the brain to exert causal effects in the world is also unknown.

However, brain's electrical activity correlating with conscious perception of a stimulus, apparently shows that it can occur *after* we respond to that stimulus, seemingly consciously. Based on this, some scientific and philosophical traditions conclude that we act non-consciously and have subsequent false memories of conscious action. This is the reason why they cast consciousness as a epiphenomenal and illusory phenomena (e.g., Daniel C. Dennett [28], *Consciousness Explained*; Wiener Norbert [20]).

Today, there is little doubt that our *volitional ability* represents the highest form of control of any mechanism or organism. After observing fantastically complex abilities of animals, such as awareness, cognition, learning, and motor control, some researchers came to the conclusion that they are the products of the mechanistic operation of their brains. It is claimed that these "mental" abilities emerge from the specific interaction between neurons, molecules, and atoms. In order to justify this, ample evidences have been gathered by evolutionary biologists, developmental psychologists and computer scientists. In addition to this, there are indications that these interactions are entirely subject to the known laws of physics and chemistry. It has almost been accepted by modern scientists and has become an established truth that, in time, machines will have all the competence and functionalities of animals.

I spite of these developments, a group of practitioners of science and technology strongly believe that while all the wonderful abilities of some animals, including consciousness and goal-directed behavior, are indeed the result of mechanistic processes, there is no way human consciousness and choice (and possibly that of some of the higher animals) can simply be the result of an essentially Newtonian physics.

If scientists are able to **genetically modify** chimpanzees so that they are endowed with such human abilities as human language ability, intelligence, and freewill, then these developments and augmentations would completely remove the fiction of the immaterial human mind and the soul. As a matter of fact these experiments need not be carried out at all, since we already have enough data in the form of healthy babies and also brain-damaged adults. They operate pretty much at the level of the higher animals. However, this similarity may not provide an irrefutable argument, yet they strongly suggest that additional neuronal circuits and connections are responsible and required for our extra capabilities. Research by the developmental and pathological psychologists and correlation between DNA and cognitive ability also provide overwhelming and convincing evidence in favor of a **naturalistic account** of consciousness and freewill

10.2.1 Quantum indeterminacy and free will

The idea of **quantum indeterminacy** (the fact that a quantum system can never predict an outcome with certainty, but only as a probability from among the various possible states) have been put forth by some proponents of quantum consciousness. This view amounts to claiming that quantum consciousness resolves the problem of whether or not humans actually have free will. The argument for this is as follows: if human consciousness is *governed by quantum physical processes*, then it is not deterministic, and humans, therefore, have free will.

10.3 Biological issues/ "quantum bio-informatics" and quantum information science

Biological Information, **Bioinformatics,** involves the integration of computers, software tools, and databases in an effort to address biological questions and biological information. The wealth of genome sequencing information has required the design of software and the use of computers to process this information.

There are two important large scale activities that use **bioinformatics**. They are genomics and proteomics. Genomics or genetics is concerned with the analysis of genomes. Scientists think about genome in terms of a complete set of DNA, RNA sequences. They code for the hereditary material that is passed on from generation to the next. The abundance of information about genome sequencing has required the design of software and the use of computers to process this information. Proteomics, on the other hand, refers to the analysis of the complete set of proteins or proteome, protein structures and various synthesis processes. Recent work in Proteomics include metabolomics, transcriptomics.

For the future of bioinformatics a key research question would be as to how to computationally compare complex biological observations, such as gene expression patterns and protein networks. Bioinformatics converts **biological observations to a model** that a computer will be able to process (or understand).

Of course Quantum Mechanics is the fundamental theory that describes the properties of subatomic particles, atoms, molecules, molecular assemblies. However, Quantum Mechanics operates on the nanometer and sub-nanometer scales. This forms the basis of fundamental life processes such as photosynthesis, respiration and vision. The fundamental claim by Quantum Mechanics is that all objects have wave-like properties and when they interact, quantum coherence describes the correlations between the physical quantities describing such objects that have a wave-like nature.

11. Conclusion

In Quantum Information Science **the physical approach** to 'Information' is found to be most appropriate approach in explaining our understanding of the consequences of representing and processing *information quantum mechanically*. Quantum Information Science is reinforced by its three pillars viz. 'Qubit', 'Superposition' and 'entanglement' and their practical and technological applications. It is facilitated by the insights of the physical properties of nature of microscopic systems at the scale of atoms and subatomic particles.

Surprisingly the physical approach to Quantum Information Science is equally significant and apt in dealing with "consciousness," "freewill" and "bioinformatics." Penrose and his collaborator, Stuart Hameroff, maintained that human intelligence is far more subtle than 'artificial intelligence' and suggested a biological analog to quantum computation involving microtubules. In neurons, microtubules (which inhabit in neurons in the brain) help control the strength of synaptic connections. In the Penrose-Hameroff theory of Orchestrated Objective Reduction, known as Orch-OR, the moments of conscious awareness are orchestrated by the microtubules in our brains, which, they believe, have the capacity to store and process information and memory. Orch OR Model and biological theories of mind are important in the area known as "bioinformatics."

Acknowledgements

I have no words to express my sincere and deep gratitude to my Editor from IntechOpen for a full chapter Review consisting of meticulous and painstaking criticisms, and insightful suggestions which enabled me to bring the paper to its present somewhat decent shape. I also cannot refrain myself from mentioning the name, Mrs. Chaitali Gupta, my wife, whose care and constant encouragement to complete the paper made it possible for me to work hard and finish it during the time of pandemic and devastations by an extremely severe cyclonic storm arising from the Arabian sea, called Tauktae, and when I was not keeping well. I must convey my sincere appreciation to Ms. Madhavi Rao for going out of her way in helping me to bringing the paper in its present form.

Author details

Amitabha Gupta
Mathematical Logic and Philosophy of Science, Indian Institute of Technology Bombay, India

*Address all correspondence to: agcg503@gmail.com

References

[1] Claude E. Shannon (1948) "A Mathematical Theory of Communication" in *Bell System Technical Journal*

[2] Floridi, Luciano (2003) "What is the Philosophy of Information?" *Metaphilosophy*, 33.1/2: 123-145. Reprinted in T.W. Bynum and J.H. Moor (eds.), 2003. *CyberPhilosophy: The Intersection of Philosophy and Computing*. Oxford – New York: Blackwell.

[3] Dennett, D. C. (1971), *Intentional Systems*, The Journal of Philosophy, Vol. 68, No. 4, pp. 87-106

[4] Searle, John R. (1997), The Mystery of Consciousness. ISBN 0-940322-06-4. pp 85–86.

[5] Searle, John R (1980): ``Minds, Brains and Programs" in Behavioral and Brain Sciences, Vol. 3. No. 3, pp. 417-458.

[6] Dretske, F. I. (1981). *Knowledge and the Flow of Information*. MIT Press, Cambridge, Massachusetts

[7] Wittgenstein, Ludwig (1953) *Philosophical Investigations* , G.E.M. Anscombe and R. Rhees (eds.), G.E.M. Anscombe (trans.), Oxford: Blackwell.

[8] Strawson, P. F. (1950), On Referring, Mind, New Series, Vol. 59, No. 235. pp. 320-344.

[9] Bayes, Thomas (1763), "An Essay Towards Solving a Problem in the Doctrine of Chances" in Philosophical Transactions of the Royal Society of London, 1763

[10] Einstein, A., B. Podolsky, and N. Rosen, (1935), "Can quantum-mechanical description of physical reality be considered complete?", *Physical Review*, 47: 777–780, DOI: https://doi.org/10.1017/CBO9780511622687.012, January 2010

[11] Carnap, Rudolf and Bar-Hillel, Yehoshua (1952) *'An Outline of a Theory of Semantic Information'*, Technical Report No. 247, Research Laboratory of Electronics, Massachusetts Institute of Technology

[12] Caves, Carlton M., Fuchs, Christopher A. and Schack, Rüdiger, (2002) Quantum probabilities as Bayesian probabilities, *Physical Review*. A 65, 022305, DOI:https://doi.org/10.1103/PhysRevA.65.022305

[13] Barwise, John. and Seligman, Jerry. (1997) *Information Flow: The Logic of Distributed Systems,* Cambridge, UK: Cambridge University Press.

[14] Landauer, Rolf (1961). Dissipation and heat generation in the computing process. IBM Journal of Research and Development, 5, 183–191.

[15] Landauer, Rolf (1991) Information is Physical, Physics Today **44**(5), https://doi.org/10.1063/1.881299

[16] Landauer, Rolf (1996) The physical nature of information, Physics Letters A, Volume 217, Issue 4, p. 188-193. DOI:10.1016/0375-9601(96)00453-7

[17] Lloyd, S. Rolf Landauer (1927-99). Head and heart of the physics of information *Nature* 400, 720 (1999). https://doi.org/10.1038/23376, DOI: https://doi.org/10.1038/23376

[18] Galindo, A and Martin-Delgado, M. A. (2002) Information and computation: Classical and quantum aspects, Reviews of Mododern Physics 74, 347, DOI: 10.1103/RevModPhys, 74.347

[19] Bennett, Charles H. *(2003)*, "Notes on Landauer's principle, Reversible Computation and Maxwell's Demon" *(PDF), Studies in History and Philosophy of Modern Physics,* ***34*** *(3): 501–510,*

arXiv:physics/0210005, Bibcode: 2003SHPMP..34..501B, doi:10.1016/S1355-2198(03)00039-X, S2CID 9648186,

[20] Wiener, Norbert (1961) Cybernetics or control and communication in the animal and the machine, Cambridge, Mass.: MIT Press

[21] Schumacher, Benjamin (1995) "Quantum coding," Physics Review. A 51, 2738–2747

[22] *Dirac, P. A. M. (1939). "A new notation for quantum mechanics". Mathematical Proceedings of the Cambridge Philosophical Society.* **35** *(3): 416–418.* Bibcode: 1939PCPS...35..416D. doi:10.1017/S0305004100021162. Also in standard text, *The Principles of Quantum Mechanics*, IV edition, Clarendon Press (1958), ISBN 978-0198520115

[23] Gupta, Amitabha (2018), *Some Aspects of the Arcane Nature of Quantum Mechanical Theory and Distinctive Forms of Realism Arising Out of It*, Journal of Indian Council of Philosophical Research, Vol. 36, issue 2, 30 Nov 2018, Pages: 295 – 338, Springer

[24] Schrödinger, E. (1935) Discussion of Probability Relations between Separated Systems, Proc. Cambridge Philos. Soc. 31, 555

[25] Penrose, Roger (1989) *The Emperor's New Mind: Concerning Computers, Minds, and the Laws of Physics*, Oxford University Press.

[26] Gupta, Amitabha (2020), *Gödel's Life and Work: before and after the Incompleteness Theorems*, to be published

[27] Penrose, Roger (1994) *Shadows of the Mind: A Search for the Missing Science of Consciousness*, Oxford University Press

[28] Dennett, Daniel C. (1991), *Consciousness Explained,* Penguin Books

Section 3

Applications

Chapter 5

Discretization, the Road to Quantum Computing?

Jesús Lacalle

Abstract

The main challenge we face in making quantum computing a reality is error control. For this reason it is necessary to study whether the hypotheses on which the threshold theorem has been proved capture all the characteristics of quantum errors. The extraordinary difficulties that we find to control quantum errors effectively together with the little progress in this endeavor, compared to the enormous effort deployed by the scientific community and by companies and governments, should make us reflect on the road map to quantum computing. In this work we analyze error control in quantum computing and suggest that discrete quantum computing models should be explored. In this sense, we present a concrete model but, above all, we propose that Quantum Physics should be taken one step further, in order to allow discretization of the quantum computing model.

Keywords: quantum computing errors, quantum threshold theorem, discrete quantum computing errors, continuous quantum computing errors, discrete quantum computing, quantum physics

1. Introduction

Quantum computing is a multidisciplinary research area with extraordinary expectations in Computer Science [1, 2]. It proposes a radical change with respect to the classical computing model, moving to a quantum one. To do this, change the basic unit of classical information, the bit, for the quantum bit or qubit:

$$\begin{aligned} &\text{Bit}: && b \in \{0,1\} \quad \text{and} \\ &\text{Qubit}: && q \in \left\{ \alpha_0|0\rangle + \alpha_1|1\rangle \;\middle|\; \alpha_0, \alpha_1 \in \mathbb{C} \text{ and } |\alpha_0|^2 + |\alpha_1|^2 = 1 \right\}. \end{aligned} \tag{1}$$

The superposition principle of Quantum Physics makes the so-called quantum parallelism possible. Working with n qubits, quantum parallelism allows 2^n operations to be performed simultaneously. However, making this advantage effective by getting algorithms faster is a difficult challenge. Another important consequence of the superposition principle is the existence of entangled quantum states. The smallest entangled state is built with 2 qubits and is called an EPR pair, because it was first proposed by Einstein, Podolsky and Rosen in 1935:

$$q = \frac{1}{\sqrt{2}}(|00\rangle + |11\rangle). \tag{2}$$

Another important feature of quantum computing is that it is a continuous computing model. Change a bit, which can only take two discrete values, for a qubit, which is a point on the 3−dimensional unit sphere centered at 0 in the real space $\mathbb{R}^4$. This fact makes quantum error control the main challenge for the feasibility of quantum computing. For this reason, one of the main research objectives in the 1990s was to solve this stumbling block. To address the problem, two fundamental tools were developed: quantum error correction codes [3–8] in combination with fault tolerant quantum computing [9–15].

The results obtained seemed to have theoretically solved the problem of quantum error control. The quantum threshold theorem or quantum fault-tolerance theorem was proved. This states that a quantum computer with a physical error rate below a certain threshold can, through application of quantum error correction schemes, suppress the logical error rate to arbitrarily low levels. Shor first proved a weak version [9] and the theorem was independently proven by the groups of Aharanov and Ben-Or [15], Knill, Laflamme and Zurek [13] and Kitaev [14].

All authors use the discrete errors introduced to define error-correcting quantum codes as a key element to prove the quantum threshold theorem. And they do it for two reasons: the constructed quantum codes allow correcting precisely those discrete errors and, even more important, any 1−qubit unitary matrix is a linear combination of those discrete errors. Indeed the discrete errors of a qubit are linear combinations of the well-known Pauli matrices:

$$I = \begin{pmatrix} 1 & 0 \\ 0 & 1 \end{pmatrix}, \ X = \begin{pmatrix} 0 & 1 \\ 1 & 0 \end{pmatrix}, \ Y = \begin{pmatrix} 0 & -i \\ i & 0 \end{pmatrix} \text{ and } Z = \begin{pmatrix} 1 & 0 \\ 0 & -1 \end{pmatrix}. \tag{3}$$

However, recent studies [16, 17] indicate that fault-tolerant quantum computing does not cover all the loopholes through which quantum errors escape, accumulating during quantum computations. Lacalle, Pozo-Coronado, Fonseca de Oliveira and Martín-Cuevas model quantum errors as random variables, integrating the essentially continuous character of quantum errors. The first two authors obtain the formula for the variance of the sum of two independent quantum errors E_1 and E_2 [18]:

$$V(E_1 + E_2) = V(E_1) + V(E_2) - \frac{V(E_1)V(E_2)}{2}. \tag{4}$$

They prove it only for isotropic errors and conjecture that it is true in the general case. The $n-$qubits are represented by points on a $(2^{n+1} - 1)-$dimensional unit sphere $\mathbb{S}$ centered at 0 in the real space $\mathbb{R}^{2^{n+1}}$. Therefore, the variance of the quantum errors of the $n-$qubits, unlike what happens in $\mathbb{R}^n$, is bounded because the corresponding sphere is a closed and bounded set. In fact the variance always belongs to the interval $[0, 4]$.

The authors establish in [16, 17] that a quantum code fixes a quantum error if, assuming that the code's correcting circuit does not introduce new errors, the code reduces the variance of the quantum error. Despite these weak requirements, the authors find two types of quantum error that are not fixed by any quantum code. Let $\mathcal{C}$ be the quantum code used, Φ the pure quantum state that the $n-$qubit should have if no error occurs, Ψ the real quantum state of the $n-$qubit generated by the quantum error and $\tilde{\Phi}$ the code state resulting from applying the code correction circuit to the state Ψ, assuming that this circuit does not introduce new errors. From the point of view of the statistical study of errors, the disturbed state Ψ is a random variable on the sphere $\mathbb{S}$. The same holds for the state $\tilde{\Phi}$ resulting from the correction, in this case on the corresponding sphere of the subspace code of $\mathcal{C}$ (since the accuracy of the correction circuit we are assuming implies that $\tilde{\Phi}$ belongs to $\mathcal{C}$). The variance of the quantum error is the expected value

$$V(\Psi) = E\left[\|\Phi - \Psi\|^2\right] \tag{5}$$

and the variance of the corrected state $V\left(\tilde{\Phi}\right) = E\left[\|\Phi - \tilde{\Phi}\|^2\right]$. Then $\mathcal{C}$ fixes the quantum error if:

$$V\left(\tilde{\Phi}\right) < V(\Psi). \tag{6}$$

The authors say in [16] that a quantum error Ψ is isotropic if its density function on the sphere $\mathbb{S}$ only depends on $\|\Phi - \Psi\|$ (θ_0, the first angle in polar coordinates). And they prove the following results:

1. If $\mathcal{C}$ detects an error the distribution of $\tilde{\Phi}$ is uniform ([16], Theorem 3).

2. $V\left(\tilde{\Phi}\right) \geq V(\Psi)$ for common probability distributions ([16], Theorem 5).

The first of the above properties indicates that if an error is detected in the code correcting circuit, all information has already been lost in computing. This result, despite being very negative from the point of view of quantum error control, is not surprising for isotropic errors.

The other type of quantum error studied by the authors in [17] is more important: qubit independent errors. They are much more difficult to analyze because they do not have as much symmetry as isotropic errors but they are errors that occur in real quantum computers. To facilitate the analysis, the authors focus on the 5−qubit quantum code because of its high symmetry and argue that the behavior of this quantum code shows a general pattern. Although these two types of errors are very different (the dimension of the support of the isotropic errors is $2^{n+1} - 1$ while that of the qubit independent errors is much smaller: $4n$), the main results are surprisingly similar. In this case the authors prove the following results:

1. If $\mathcal{C}$ detects an error the distribution of $\tilde{\Phi}$ has central symmetry ([17], Theorem 4.2) and its variance is maximum ([17], Lemma 4.2).

2. $V\left(\tilde{\Phi}\right) \geq V(\Psi)$ for common probability distributions ([17], Theorem 4.4).

Note that the second property is the same for both types of quantum error. And, as regards the first, there is not much difference between a uniform distribution on a sphere and a centrally symmetric distribution, if they both approximate a point Φ on the sphere. Therefore, the results for both types of quantum error are similar and this fact is very striking.

Some reviewers have questioned the result of [17] for not considering that quantum states can be multiplied by a phase without physically changing their state. However, the authors of this work introduce the quantum variance that considers this fact,

$$V_q(\Psi) = E\left[\min_{\phi}\left(\|\Psi - e^{i\phi}\Phi\|^2\right)\right], \tag{7}$$

and relate it to the most common error measure in quantum computing, fidelity $F(\Psi)$:

$$1 - \frac{V_q(\Psi)}{2} \le F(\Psi) \le \sqrt{1 - \frac{V_q(\Psi)}{2}}. \quad (8)$$

These inequalities show that quantum variance and fidelity are essentially equivalent, since when quantum variance tends to 0, fidelity tends to 1 and, conversely, when fidelity tends to 1, quantum variance tends to 0. Of the three measures, the variance is the only one that allows to complete the complicated calculations performed in [17]. Furthermore, the authors state that the variance and the quantum variance have similar behaviors for continuous quantum computing errors. Indeed, let $\Phi = |0\rangle$ be a qubit and suppose that Φ is changed by error becoming the state $\Psi = W\Phi$, where W is the error operator given by Formula (5) in [17] whose density function $f(\theta_0)$ only depends on the angle θ_0. Then:

$$\begin{aligned}\Psi = &(\cos(\theta_0) + i \sin(\theta_0) \cos(\theta_1)) |0\rangle + \\ &(\sin(\theta_0) \sin(\theta_1) \cos(\theta_2) + i \sin(\theta_0) \sin(\theta_1) \sin(\theta_2)) |1\rangle\end{aligned} \quad (9)$$

and, taking into account that

$$\min_{\phi} (\|\Psi - e^{i\phi}\Phi\|^2) = 2 - 2|\langle\Psi|\Phi\rangle| \quad (10)$$

and the Eq. (5) we obtain:

$$V_q(X) = 2 - 4\pi\int_0^{\pi}\left(1 - \frac{\cos^2(\theta_0)}{2\sin(\theta_0)}\log\left(\frac{1 - \sin(\theta_0)}{1 + \sin(\theta_0)}\right)\right)\cdot f(\theta_0)\sin^2(\theta_0)d\theta_0 \text{ and}$$

$$V(X) = 2 - 4\pi\int_0^{\pi} 2\cos(\theta_0)\cdot f(\theta_0)\sin^2(\theta_0)d\theta_0.$$

We observe that the difference between the quantum variance and the variance are the weight functions of $f(\theta_0)\sin^2(\theta_0)$ in the integral and that they have a similar behavior for small errors, that is, for concentrated density functions $f(\theta_0)$ around $\theta_0 = 0$ (see **Figure 1**).

Even for large errors, for example a uniform distribution function $f = \frac{1}{2\pi^2}$, we have comparable values of the quantum variance and the variance:

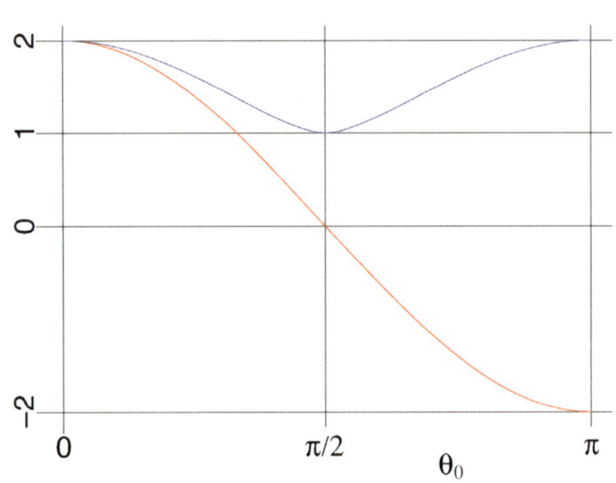

Figure 1.
Weight functions for quantum variance (red) and variance (blue).

$$V_q(\Psi) = \frac{2}{3} \text{ and } V(\Psi) = 2. \tag{11}$$

In [19], the study of isotropic errors is extended by analyzing the capacity of quantum codes to improve fidelity, and similar results to those presented in [16] are obtained: quantum codes do not improve the fidelity of uncoded quantum states for this type of error.

The results presented in [16, 17, 19] remind us that the quantum computing model is continuous and that the treatment of continuous quantum errors has many subtleties and it is an extraordinarily difficult challenge. Right now we are at a crossroads: extend fault-tolerant quantum computing to error models that include continuous errors or search for a discrete model of quantum computing that allows easier error control. The first road presents formidable difficulties: the fault-tolerant quantum architecture is based exclusively on discrete quantum errors and there is no analogical (continuous) system in the world comparable in complexity to a computer. The second one includes two processes: defining a discrete quantum computing model and finding a quantum system that allows the model to be implemented. It is difficult to know which of the two approaches will lead us to real quantum computing and, for this reason, both should be explored. In this work we study the second one.

A discrete quantum computing model has already been published [20] and, as far as we know, it is the first. In this work Gatti and Lacalle present a discrete quantum computing model based on the following basic requirements:

1. It describes real states in Quantum Physics.

2. It preserves the main characteristics of quantum states: superposition, parallelism and entanglement.

3. It allows to approximate general quantum states.

4. It contains simple quantum states.

Of all the possible sets of discrete quantum states, there is one that, fulfilling the first three properties, is the most outstanding in terms of simplicity of the states. It is the set of Gaussian coordinate states, which includes all the quantum states whose coordinates in the computation base, except for a normalization factor $\sqrt{2}^{-k}$, belong to the ring of Gaussian integers:

$$\mathbb{Z}[i] = \{a + bi \mid a, b \in \mathbb{Z}\}. \tag{12}$$

To define the model they also need to introduce a set of quantum gates that verify the following properties: it contains quantum gates that transform discrete states into discrete states, and it generates all discrete quantum states. And they includes two elementary quantum gates that verify the above properties, H and G. The Hadamard gate H allows superposition, while the other one, G, is a 3−qubit quantum gate. Two of them are control qubits, while the third is the target. If the control qubits are in state $|1\rangle$, then the quantum gate V is applied to the third qubit:

$$V = \begin{pmatrix} 1 & 0 \\ 0 & i \end{pmatrix}. \tag{13}$$

This quantum gate allows the construction of all Gaussian coordinate states (discrete states) and it is because of this that they call it G.

This model of discrete quantum computing is related to Number Theory since discrete quantum states

$$\Phi = \frac{(a_0 + ib_0, a_1 + ib_1, \ldots, a_{2^n-1} + ib_{2^n-1})}{\sqrt{2}^k} \tag{14}$$

must verify the following diophantine equation:

$$a_0^2 + b_0^2 + a_1^2 + b_1^2 + \cdots + a_{2^n-1}^2 + b_{2^n-1}^2 = 2^k, \tag{15}$$

where $k \in \mathbb{N}$ and $a_0, b_0, \ldots, a_{2^n-1}, b_{2^n-1} \in \mathbb{Z}$. The above equation establishes deep connections between the discrete quantum computing model and Lagrange's four-square theorem. The same authors analyze this relationship in [21].

However, we must go one step further with the model of discrete quantum computing, so do not have the same error handling problem again. We need the discrete quantum states to have a basin of attraction associated with them so that any state that falls inside is automatically self-correcting, transforming into the discrete state. This process is used in the manufacture of hardware for classic computers with enormously satisfactory results.

However, Quantum Physics does not allow the application of this process. First of all, self-correction is not a one-to-one transformation and therefore cannot be unitary. And secondly, it cannot be the result of a quantum measurement either because the probability that the result was not the associated discrete state would be greater than zero. Consequently, we need Quantum Physics to go one step further to have the control that discrete quantum computing requires. Is this possible? We believe that this question should have an affirmative answer if the following one does: Is quantum computing possible?

In the following sections we develop further the ideas presented in this introduction.

2. Overview of quantum error control

Today's quantum error control has two essential components: quantum error correction codes [3–8] and fault-tolerant quantum computing [9–15]. There are textbooks on this subject, such as Gaitan's [22].

2.1 Quantum error correcting codes

Calderbank and Shor [3] and Steane [4] discovered an important class of quantum error correcting codes. The Calderbank-Shor-Steane (CSS) codes are constructed from two classical binary codes. Another approach to the subject originated the quantum stabilizer codes [5–8]. However, to better understand the role of quantum codes in correcting errors, a general description of them is more useful, without going into the detail of their internal structure.

An quantum error correcting code of dimension $[n, m]$ is a subspace $\mathcal{C}$ of dimension $d' = 2^m$ in the $n-$qubit space $\mathcal{H}^n$, whose dimension is $d = 2^n$. The $\mathcal{C}$ quantum code encoding function is a unitary operator C that satisfies the following properties:

$$C : \mathcal{H}^m \otimes \mathcal{H}^{n-m} \to \mathcal{H}^n \text{ and } \mathcal{C} = C(\mathcal{H}^m \otimes |0\rangle). \tag{16}$$

The $\mathcal{C}$ code fixes $d'' = 2^{n-m}$ discrete errors: $E_0,\ E_1,\ \ldots,\ E_{d''-1}$. Since the identity I should be among these unitary operators, we assume that $E_0 = I$. This process of discretization of errors allows to correct any of them if the subspaces $S_s = E_s(\mathcal{C})$, $0 \le s < d''$, satisfy the following property:

$$\mathcal{H}^n = S_0 \perp S_1 \cdots \perp S_{d''-1}. \tag{17}$$

That is, $\mathcal{H}^n$ is the orthogonal direct sum of said subspaces. Note also that $S_0 = E_0(\mathcal{C}) = I(\mathcal{C}) = \mathcal{C}$. In the stabilized code formalism, the code $\mathcal{C}$ is the subspace of fixed states of an abelian subgroup of the Pauli group $\mathcal{P}_n = \{\pm 1, \pm i\} \times \{I, X, Z, Y\}^n$ and discrete errors are operators of $\mathcal{P}_n$ that anti-commute with any of the subgroup generators, except for the identity operator E_0. If Formula (17) holds, the code is non-degenerate.

Suppose that a coded state Φ is changed by error, becoming the state Ψ. The initial state is a code state, that is, $\Phi \in S_0$, while the final state in general is not, that is, $\Psi \notin S_0$. If the disturbed state belongs to the subspace $W_\Phi = L(E_0\Phi, \dots, E_{d''-1}\Phi)$, that is, if it is of the form

$$\Psi = \alpha_0 E_0 \Phi + \cdots + \alpha_{d''-1} E_{d''-1} \Phi \quad \text{with} \quad |\alpha_0|^2 + \cdots + |\alpha_{d''-1}|^2 = 1, \tag{18}$$

then the quantum code allows us to retrieve the initial state Φ. To achieve this, we measure Ψ with respect to the orthogonal decomposition of the Formula (17). The result will be $\frac{\alpha_s}{|\alpha_s|} E_s \Phi$ for a value s between 0 and $d'' - 1$. The value of s is called syndrome and allows us to identify the discrete error that the quantum measurement indicates. Then, applying the quantum operator E_s^{-1} we obtain $\frac{\alpha_s}{|\alpha_s|}\Phi$. This state is not exactly Φ but, differing only in a phase factor, both states are indistinguishable from the point of view of Quantum Mechanics. Therefore, the code has fixed the error.

An error that does not satisfy Formula (18), that is, it does not belong to W_Φ, cannot be fixed exactly. For example, if Ψ belongs to the code subspace $\mathcal{C}$, the error cannot be fixed at all since, being a code state, it is assumed that it has not been disturbed. Therefore it is important to analyze the limitation in the correction capacity of an arbitrary code, assuming that the code correction circuit does not introduce new errors.

Finally, we want to highlight that discrete errors can be chosen so that, for example, all errors affecting a single qubit are fixed. The best code with this feature that encodes one qubit is the 5-qubit quantum code [23, 24]. This code is optimal in the sense that no code with less than 5 qubits can fix all the errors of one qubit.

2.2 Fault-tolerant quantum computing

Fault-tolerant quantum computing was proposed with the aim of proving the quantum threshold theorem or quantum fault-tolerance theorem: a quantum computer with a physical error rate below a certain threshold can, through application of quantum error correction schemes, suppress the logical error rate to arbitrarily low levels. Shor first proved a weak version [9] and the theorem was independently proven by the groups of Aharanov and Ben-Or [15], Knill, Laflamme and Zurek [13] and Kitaev [14].

The essential elements of fault-tolerant quantum computing [9, 13, 15] are as follows: the encoding of each of the qubits with quantum error-correcting codes, the use of fault-tolerant quantum gates, the application of quantum gates on coded qubits (encoded operations) and the concatenation of quantum error-correcting codes.

Another essential element for the proof of the quantum threshold theorem is the quantum error model used. Shor [9] assumes that there is no decoherence error and considers that in a quantum gate an error occurs with probability p and that the errors corresponding to different qubits are independent. Therefore the probability that errors will occur in k qubits simultaneously is:

$$\text{Prob}(k \text{ errors}) = \binom{n}{k}(1-p)^{n-k}p^k. \tag{19}$$

Knill, Laflamme and Zurek [13] and Aharanov and Ben-Or [15] consider both decoherence errors and errors in quantum gates and also assume the independence of errors on different qubits. The first [13] analyze quasi-independent and monotonic errors with error strength p and bound C: the total strength of the summands for which at least a given k many error locations have failed is at most Cp^k. Aharanov and Ben-Or [15] use density matrices and model the error in a qubit as follows:

$$(1-p)I + pE. \tag{20}$$

In all cases, the parameter p can be considered as the probability that an error occurs in a qubit and therefore the probability that k errors coincide in different qubits will be proportional to p^k. This consideration is key in proving the quantum threshold theorem and as such it appears in Gaitan's textbook [22] (see for example Table 1.1 on page 38). The errors associated with p are arbitrary and include what Shor calls "fast" errors and also "slow" errors. In particular they include the errors described by the Pauli matrices (3). This error model is the discretized quantum error model or the stochastic quantum error model.

The discretized quantum error model together with the concatenation of error-correcting quantum codes are the key elements in the proof of the quantum threshold theorem. The effect of the conjugation of both is as follows (see for example Figure 6 in [13]):

	Uncoded	Coded once	Coded twice
Number of qubits	1	7	49
Error probability	p	p^2	p^4

(21)

where we have used the 7−qubit CSS code. In each encoding the error in a qubit is fixed by the code and only errors of order 2 or greater remain. This scheme makes the error small, since p^k tends to zero if k grows.

But this approach cannot be used in all cases, for example for the decoherence error, since in this case the reality is different: the probability of errors occurring in all qubits is 1, although on the other hand the errors with high probability are small. In this situation the correcting code cannot handle a simultaneous error in all qubits and neither can it correct the "lower order" errors. Here is the essential difference between the discrete error model and the continuous one. The discrete error model does not fit this situation, in which small errors are not controlled and, after the application of the code correction circuit, become undetectable (because the resulting state belongs to the subspace code) and accumulate during computation.

Another key to fault-tolerant quantum computing is to avoid quantum gates that act on two qubits belonging to the same quantum code instance (implementation of fault-tolerant quantum gates for the used quantum code). In this way, the imprecision of the quantum gates only introduces error in at most one qubit of each instance of the quantum code. However, the error in 2−qubit quantum gates is not reduced to an error in each of the qubits. It also generates an error that affects both qubits simultaneously (entangled error) and the code instances to which the two qubits belong are not designed to tackle it.

The use of an instance of an error correcting quantum code of dimension $[n, 1]$ on each of the qubits of a quantum circuit (algorithm) produces two additional effects to consider. First, this multiplies the number of qubits in the circuit by n. As a consequence, the decoherence per unit of time that occurs in the circuit is multiplied by n. Second, the number of gates in the circuit is multiplied by at least $n(n+1)$. Each encoded quantum gate requires a minimum of n quantum gates and, after each one of them, the code correction circuit must be applied, that is, at least another n quantum gates or measurements are needed. The effect of this increasing number of quantum gates is that the imprecision errors are multiplied by $n(n+1)$. A total of at least n^2 of these quantum gates and measurements correspond to the correction circuits and are therefore not protected. This fact remains even if we concatenate quantum codes in the last application of the error correcting code. If the number of quantum gates in an algorithm is n and the error correcting code is concatenated k times, the final number of gates is at least n^{2^k}. Then, the ratio of quantum gates not protected from imprecision errors is at least

$$1 - \frac{1}{n^{2^{k-1}}}. \tag{22}$$

Finally, it should be noted that the use of quantum codes produces an additional increase in decoherence by increasing the execution time of the algorithms.

Despite the difficulties raised above for the effective control of quantum errors, the discrete quantum error model or stochastic quantum error model allows the proof of the quantum threshold theorem. But unfortunately this model of quantum computing errors does not allow a realistic analysis of continuous quantum computing errors. These break the golden rule of error correction: all small errors must be corrected. The road of fault-tolerant quantum computing goes through including continuous errors in the quantum threshold theorem. This is a huge challenge and for this reason it is interesting to investigate other possible roads.

3. Discrete quantum computing

We are interested in discrete quantum computing because it could lead us to a quantum computing where error control was an easier challenge. In the literature there are some works on discrete quantum computing. They generally intend to simplify or better understand the quantum model: introducing modal concepts and finite fields for the representation of quantum amplitudes [25–29], using discretization for the design of algorithms [30], relating the structures of computation and the foundations of physics [31–38] and studying universal sets of discrete quantum gates [39–43].

As we have already commented in the Introduction, a discrete quantum computing model has already been published [20]. It is a model in which discretization is applied both to quantum states and to quantum gates and that aims to become independent from the standard quantum model (continuous model) and even, if possible, from continuous hardware (Quantum Physics). The presented discrete quantum computing model is based on the following basic requirements:

1. It describes real states in Quantum Physics.

2. It preserves the main characteristics of quantum states: superposition, parallelism and entanglement.

3. It allows to approximate general quantum states.

4. It contains simple quantum states.

Of all the possible sets of discrete quantum states, there is one that, fulfilling the first three properties, is the most outstanding in terms of simplicity of the states. It is the set of Gaussian coordinate states, which includes all the quantum states whose coordinates in the computation base, except for a normalization factor $\sqrt{2}^{-k}$, belong to the ring of Gaussian integers:

$$\mathbb{Z}[i] = \{a + bi \mid a, b \in \mathbb{Z}\}. \tag{23}$$

To define the model the authors introduce a set of elementary quantum gates that verify the following properties: it contains quantum gates that transform discrete states into discrete states, and it generates all discrete quantum states. This set includes two quantum gates that verify the above properties, H and G. The Hadamard gate H allows superposition, while the other one, G, is a 3−qubit quantum gate. Two of them are control qubits, while the third one is the target. If the control qubits are in state $|1\rangle$, then the quantum gate V is applied to the third qubit:

$$V = \begin{pmatrix} 1 & 0 \\ 0 & i \end{pmatrix}. \tag{24}$$

This model of discrete quantum computing is related to Number Theory since discrete quantum states

$$\Phi = \frac{(a_0 + ib_0, a_1 + ib_1, \ldots, a_{2^n-1} + ib_{2^n-1})}{\sqrt{2}^k} \tag{25}$$

must verify the following diophantine equation:

$$a_0^2 + b_0^2 + a_1^2 + b_1^2 + \cdots + a_{2^n-1}^2 + b_{2^n-1}^2 = 2^k, \tag{26}$$

where $k \in \mathbb{N}$ and $a_0, b_0, a_1, b_1, \ldots, a_{2^n-1}, b_{2^n-1} \in \mathbb{Z}$.

As we will see in the next subsection, the level of a discrete state is defined as the lowest natural number k for which the previous diophantine Eq. (26) holds. The superposition principle of Quantum Physics is satisfied in the following case: Given orthogonal discrete states $\Phi_0, \Phi_1, \ldots, \Phi_{j-1}$ belonging to levels $k_0, k_1, \ldots, k_{j-1}$ respectively, then the following linear combinations are also discrete quantum states:

$$\Phi = \frac{(c_0 + id_0)}{\sqrt{2^{k'_0}}}\Phi_0 + \frac{(c_1 + id_1)}{\sqrt{2^{k'_1}}}\Phi_1 + \cdots + \frac{(c_{j-1} + id_{j-1})}{\sqrt{2^{k'_{j-1}}}}\Phi_{j-1} \tag{27}$$

where $k'_0,\ k'_1,\ \ldots,\ k'_{j-1} \in \mathbb{N}$, $k_0 + k'_0, k_1 + k'_1,, k_{j-1} + k'_{j-1}$ have the same parity, $c_0, d_0, c_1, d_1, \ldots, c_{j-1}, d_{j-1} \in \mathbb{Z}$ and

$$\frac{c_0^2 + d_0^2}{2^{k'_0}} + \frac{c_1^2 + d_1^2}{2^{k'_1}} + \cdots + \frac{c_{j-1}^2 + d_{j-1}^2}{2^{k'_{j-1}}} = 1. \tag{28}$$

The superposition principle is also satisfied for non-orthogonal discrete states. For example for the following two discrete states of level 4:

$$\begin{aligned}\Phi_0 &= \frac{1}{4}(1+i, 1+2i, 0, 3)\\ \Phi_1 &= \frac{1}{4}(1+i, 0, 1+2i, 3)\end{aligned} \qquad \Phi = \frac{5+9i}{8}\Phi_0 - \frac{3+9i}{8}\Phi_1. \tag{29}$$

Discrete state Φ has level 10, result of the sum of the levels of states Φ_0 and Φ_1, 4, and of coefficients used in the combination, 6.

3.1 Discrete quantum states

The quantum gates H and G, along with two auxiliary qubit (ancilla qubits), allow to perform a wide set of operations, for example, any permutation of the states of the computational base $\mathcal{B}$ and adding a factor -1, i or $-i$ to any subset of coordinates of an n-qubit, with respect to the computational base $\mathcal{B}$, where:

$$\begin{aligned}\mathcal{B} &= [|0\rangle, |1\rangle, |2\rangle, |3\rangle, |4\rangle, |5\rangle, |6\rangle, |7\rangle, |8\rangle, \dots, |2^n-1\rangle] \text{ or}\\ \mathcal{B} &= [|0\cdots00\rangle, |0\cdots01\rangle, |0\cdots10\rangle, |0\cdots11\rangle, \dots, |1\cdots11\rangle].\end{aligned} \tag{30}$$

They also allow obtaining other quantum gates that are commonly used: X, $\Lambda X = Cnot$, $\Lambda^2 X = Toffoli$, Z, ΛZ, $\Lambda^2 Z$, V and ΛV.

The set of discrete quantum states $\mathcal{E}$ is defined as follows: $\mathcal{E}$ is the smallest set of quantum states which contains the computational base and is invariant under the application of the conforming gates H and G. As a consequence of the properties of H and G discussed above, the set $\mathcal{E}$ is also invariant by any permutation of coordinates and by the addition of a factor -1, i or $-i$ to any subset of coordinates.

The conforming quantum gates H and G have been chosen in order to generate exactly the states whose coordinates are Gaussian integers (except for a normalization factor of the form $\sqrt{2}^{-k}$ where $k \in \mathbb{N}$) that is, elements of the set $\mathbb{Z}[i]$ defined in Formula (23).

The set of Gaussian coordinate states E is defined by the following property: a quantum state $\Phi \in E$ if and only if there exists $k \in \mathbb{N}$ such that $\sqrt{2}^k \Phi \in \mathbb{Z}[i]^{2^n}$. And, as we have already commented before, the set of discrete states $\mathcal{E}$ and the set of Gaussian coordinate states E are the same. Consequently every discrete state must verify the Eq. (25), for a certain value $k \in \mathbb{N}$, and its coordinates without the normalization factor the diophantine Eq. (26).

Discrete states are classified by levels. We say that a discrete state Φ is at level $k \in \mathbb{N}$ if k is the smallest natural number for which it is verified that $\sqrt{2}^k \Phi \in \mathbb{Z}[i]^{2^n}$. From Eq. (25) it is concluded that there is a one-to-one relationship between the discrete states and the integer solutions of the Eq. (26) in which at least one component (real or imaginary part) of one coordinate is odd.

Given $k \in \mathbb{N}$, we call E_k to the set of discrete states of level k. These sets verify the following properties: for all $k \in \mathbb{N}$ E_k is finite, in fact its size is bounded by the number of solutions of the diophantine Eq. (26); and for all $k_1, k_2 \in \mathbb{N}$, $k_1 \neq k_2$, it holds $E_{k_1} \cap E_{k_2} = \emptyset$.

Given a number $k \in \mathbb{N}$, the set of discrete states with a level less than or equal to k, $E_{\leq k}$, allows us to approximate a general quantum state with a precision of the order of $\sqrt{2}^{-k}$. In this sense, the set of discrete states E allows us to approximate general quantum states and, as the level of the discrete states increases, the approximation is more precise. Finding the best approximation of a general quantum state through a discrete state in $E_{\leq k}$, $k \geq 0$, is a natural problem that allows us to relate discrete quantum computing with quantum computing. This problem is also related to Number Theory because the discrete states must verify the diophantine Eq. (26).

In discrete quantum computing, the parity and the parity pattern of the coordinates are important. Given a coordinate $a + ib \in \mathbb{Z}[i]$ these concepts are defined as follows:

$$\begin{aligned} \mathrm{P}(a+ib) &= a+b \mod 2 \text{ and} \\ \mathrm{PP}(a+ib) &= (a \mod 2, b \mod 2). \end{aligned} \tag{31}$$

From formula (26) it is easy to deduce that the number of coordinates with parity 1 in a discrete state of level $k \geq 1$ is even.

The proof that the set of discrete states $\mathcal{E}$ is the same as the set of Gaussian coordinate states E illustrates well the structure of these sets and uses as key elements the concepts introduced above. The non-trivial part of this proof consists of giving a procedure (algorithm) to construct a state of E starting from a vector of the computational base, $|0\rangle$ for example, and applying the quantum gates H and G repeatedly. Gate H changes the level of all discrete state, most of the time increasing it by 1. But they also reduce by 1 the level of the states that we call "reducible". For example, the gate H applied to the nth-qubit, H_n, produces the following change in the discrete quantum state:

$$\begin{aligned} &\frac{1}{\sqrt{2}^k}(a_0 + ib_0, a_1 + ib_1, \ldots) \rightarrow \\ &\frac{1}{\sqrt{2}^{k+1}}((a_0 + a_1) + i(b_0 + b_1), (a_0 - a_1) + i(b_0 - b_1), \ldots). \end{aligned} \tag{32}$$

Therefore, for the state to be reducible, all the coordinates of the state resulting from the application of H_n must be multiples of 2. In this case, the initial increment by 1 of the discrete state level becomes a decrement by 1, by dividing the coordinates by 2. This division by 2 is compensated by multiplying the normalization factor $\sqrt{2}^{-(k+1)}$ by 2, that is, reducing its exponent by 2. Consequently, a state is reducible by applying H_n if its coordinates, taken two by two, have the same parity pattern:

$$\begin{aligned} &\text{Pattern } (0,0): && (\mathit{even}, \mathit{even}) && - && (\mathit{even}, \mathit{even}), \\ &\text{Pattern } (0,1): && (\mathit{even}, \mathit{odd}) && - && (\mathit{even}, \mathit{odd}), \\ &\text{Pattern } (1,0): && (\mathit{odd}, \mathit{even}) && - && (\mathit{odd}, \mathit{even}), \\ &\text{Pattern } (1,1): && (\mathit{odd}, \mathit{odd}) && - && (\mathit{odd}, \mathit{odd}). \end{aligned} \tag{33}$$

The proof starts from a discrete state of level $k \in \mathbb{N}$ and, applying the quantum gates H and G, its level is reduced, one by one, to level 0 and, once this is done, it is transformed into a state of the computational base. Then the construction of the state consists of writing this product of quantum gates in reverse order and substitute G for its inverse G^3. The keys of the proof are as follows. First, all the coordinates with the parity pattern $(0,1)$ are multiplied by i, so that all coordinates with parity 1 have the parity pattern $(1,0)$. Secondly, the coordinates are permuted so that the parity patterns $(1,0)$ appear at the end of the vector and, just before, the largest possible even number of patterns $(1,1)$ and the largest possible even number of patterns $(0,0)$.

If all the coordinates are already placed, the state is reducible. Otherwise the first two coordinates will have parity patterns $(0,0)$ and $(1,1)$ and the application of the quantum gate

$$R = V_1 H_n V_1 H_n, \tag{34}$$

where V_1 multiplies the second coordinate by i and H_n is the application of the quantum gate H to the last qubit, will solve the problem:

$$R\Phi = \frac{1}{\sqrt{2}^k}\left(\frac{a_0 - b_0 + a_1 + b_1}{2} + i\frac{a_0 + b_0 - a_1 + b_1}{2},\right.$$
$$\left.\frac{a_0 - b_0 - a_1 - b_1}{2} + i\frac{a_0 + b_0 + a_1 - b_1}{2}, a_2 + ib_2, \dots\right). \tag{35}$$

The quantum gate R plays an important role in discrete quantum computing. It modifies (rotates) the parity patterns of the first two coordinates of the n-qubit as shown in **Figure 2**.

3.2 Discrete quantum gates

The introduced discrete quantum computing model satisfies some properties that the authors did not expect to hold. They define discrete quantum gates as the quantum gates that leave the set of discrete states invariant. This means that a quantum gate is discrete if applying it to any discrete state produces another discrete state as a result.

Discrete quantum gates are characterized by a simple property: a quantum gate is discrete if and only if the columns of its matrix, with respect to the computational base, are discrete states with levels of the same parity. This characterization is also fulfilled by substituting the columns of the matrix for the rows, since the matrix is unitary.

The number of discrete gates of one-qubit is finite because the number of discrete states of one-qubit is also finite: 8 discrete states of level 0, 24 of level 1, 16 of level 2 and none of level greater than or equal to 3. In this case all discrete gates can be generated from H and G.

Like discrete states, discrete gates are classified by levels. The level of a discrete gate is defined as the highest of the levels of its columns, considered as discrete

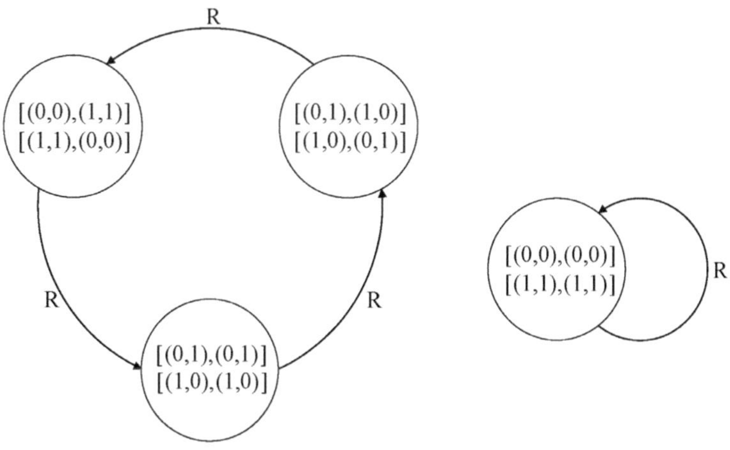

Figure 2.
Rotation of the parity patterns by the quantum gate R.

states. Obviously if we defined the level of a discrete gate using the rows instead of the columns, the result would be the same.

To proof that a discrete gate can be obtained as a product of gates H and G, it is enough to show that its level can be reduced, one by one, by left and right multiplying by these gates. This is possible only if we can make the discrete states of all its columns simultaneously reducible. And this surprisingly is possible!

Gatti and Lacalle prove it for discrete two-qubit quantum gates and conjecture that the result is true for any number of qubits. To do this, they generalize the properties of the parity patterns already introduced to the discrete gates (see **Figure 3**). They introduce the following concepts:

1. Simple match: Given two columns of a discrete gate, we will say that there is a simple match, when there exist elements in both columns, corresponding to the same row, with the real parts or the imaginary parts both odd.

2. Cross match: Given two columns of a discrete gate, we will say that there is a cross match, when there exist elements in both columns, corresponding to the same row, with the real part of one and the imaginary part of the other both odd.

From this definition and taking into account that the columns of a discrete gate are orthogonal discrete states, we can observe:

1. The number of odd elements in any column of a discrete gate is even.

2. Given two columns of a discrete gate, the number of simple matches and the number of cross matches are even.

We remark that every result about the columns of a quantum gate is also valid for the rows, since the matrix is unitary.

As it happened with the quantum states, we need to appeal to the gates R and R^t (transpose of R), which will act on the left and on the right, respectively. The gate R^t also produces a rotation of the coordinate parity patterns, analogously to the way R does (see **Figure 2**). However in this case the rotation is in the opposite direction.

The proof that discrete two-qubit quantum gates can be generated from gates H and G is much more technical than that described for discrete states. The parity constraints of the rows and columns of the discrete gates, derived from their unitarity, are sufficient tools to complete the proof. Readers interested in the details of this demonstration can refer to the original article [20]. The techniques used in the proof do not generalize for discrete gates of more than two qubits, but authors believe that the result is true in general.

Conjecture 1. For all $n \geq 3$ every dicrete n-qubit quantum gate can be decomposed into a product of H and G quantum gates.

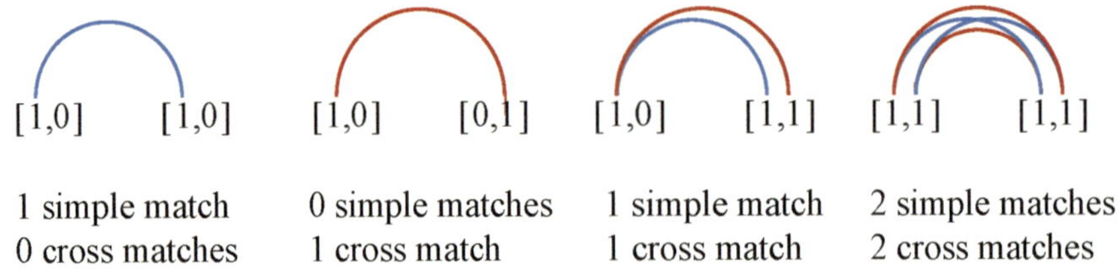

Figure 3.
Odd coordinate component matches.

4. Discrete quantum computing and Lagrange's four-square theorem

Conjecture 1 can be generalized as follows.

Conjecture 2. Given a set of $n-$qubit discrete states of levels of the same parity and orthogonal two by two, it is possible to build all of them simultaneously (applying a given circuit to different states of the computational base), using the conforming gates H and G.

Observe that the conjecture also makes sense for $2-$qubits, since in the previous subsection it has only been proved for sets of 4 discrete states. The conjecture is also interesting in the non-discrete case, since it asks about the possibility of simultaneously constructing up to 2^n quantum states simultaneously. In this case the conjecture is obviously true. Simply complete the orthonormal base, for example using the Gram-Schmidt method, and decompose the resulting unitary matrix into product of basic quantum gates. Therefore, it makes sense to ask if it is in the case of discrete quantum computing.

Before continuing, let us relax the discrete state level definition given in the previous section to any value of k for which the discrete state verifies Eq. (26). We will call these values *widespread levels*. Note that if k is a widespread level of a discrete state then $k+2$ is also. Then, a discrete state has widespread level k if and only if it is of the form k_0+2j, where k_0 is the level of the discrete state and j a natural number. This property allows to write all discrete states (with levels of the same parity) at the same widespread level.

Let us see that, somehow, building a set of orthogonal discrete states is equivalent to completing the set to an orthonormal base. For this reason we will focus in the following problem:

Problem 1. Given a natural number k and $\Psi_1, \ldots, \Psi_j$ $n-$qubit discrete states with widespread level k, $1 \le j < 2^n$, such that $\langle \Psi_i | \Psi_m \rangle = 0$ for all $1 \le i < m \le j$, then is there an $n-$qubit discrete state with widespread level k, Ψ, such that $\langle \Psi_i | \Psi \rangle = 0$ for all $1 \le i \le j$?

Considering that every discrete $2-$qubit quantum gate can be built from gates H and G, the following can be easily proved: for $2-$qubits Conjecture 2 is true if and only if Problem 1 has an affirmative answer. Then the resolution of Problem 1 would allow us to build bases with special characteristics and it would help us to demonstrate the conjecture that any $n-$qubit discrete gate, with $n \ge 3$, can be generated from quantum gates H and G.

The fact that establishes the connection between discrete quantum computing and Lagrange's four-square theorem is that the discrete states have to satisfy Eq. (26). Lagrange's four-square theorem [44] says that every natural number is a sum of four squared integer numbers and, consequently, guarantees that there exist discrete states for any level $k \ge 0$ and for any number of qubits $n \ge 1$.

Problem 1 is an orthogonal version of Lagrange's four-square theorem, i.e. the discrete state Ψ must verify the Diophantine Eq. (26) and the following orthogonality conditions:

$$\langle \Psi_i | \Psi \rangle = 0 \quad \text{for all} \quad 1 \le \mathrm{i} \le \mathrm{j}. \tag{36}$$

Note that given a value of k, if the Eq. (26) has a solution for a $1-$qubit, then it has a solution for every number of qubits $n \ge 2$. Nevertheless, this generalization is not necessarily true for the Problem 1, because of orthogonality conditions. Therefore the problem has its own entity for any number of qubits n.

Problem 1 turns out to be a difficult question in Number Theory and has deep implications. For this reason we begin with the following simplification that most resembles Lagrange's four-square problem: $n=2$, integers as coordinates instead of

Gaussian integers and normalization factor $\sqrt{p}$, being p a prime number, instead of $\sqrt{2^k}$.

Problem 2. Given a prime number p and $v_1, \ldots, v_k \in \mathbb{Z}^4$, $1 \le k \le 3$, such that $\|v_i\|^2 = p$ for all $1 \le i \le k$ and $\langle v_i | v_j \rangle = 0$ for all $1 \le i < j \le k$, then is there a vector $v = (x_1, x_2, x_3, x_4) \in \mathbb{Z}^4$ such that $\langle v_i | v \rangle = 0$ for all $1 \le i \le k$ and $\|v\|^2 = x_1^2 + x_2^2 + x_3^2 + x_4^2 = p$?

Given a natural number $1 \le k \le 4$ and a set of vectors $v_1, \ldots, v_k \in \mathbb{Z}^4$ such that $\|v_i\|^2 = p$ for all $1 \le i \le k$ and $\langle v_i | v_j \rangle = 0$ for all $1 \le i < j \le k$, we will say that $S = \{v_1, \ldots, v_k\}$ is a p*–orthonormal system* and, if $k = 4$, that S is a p*–orthonormal base*.

Given a p*–orthonormal system* S, we will call *support of* S, $supp(S)$, to $\{i \mid \exists j \text{ \{such that the\} } i\text{-\{coordinate of\} } v_j \neq 0\}$ and we will say that $|supp(S)|$ is the *support size of* S.

In this context, the problem we are dealing with (Problem 2) is stated as follows: given a prime number p and a p–orthonormal system $S = \{v_1, \ldots, v_k\}$, $1 \le k \le 3$, prove that there exists $v \in \mathbb{Z}^4$ such that $\langle v_i | v \rangle = 0$ for all $1 \le i \le k$ and $\|v\|^2 = p$.

To prove the result, the authors consider four cases. Three of them are solved with basic linear algebra techniques. However the fourth case is much more difficult, and requires the use of lattices and some Number Theory results.

Case 1: one vector p–orthonormal systems.

If the p–orthonormal system S has a single vector $v_1 = (x_1, x_2, x_3, x_4)$, the solution (valid for all $p \ge 1$) is trivial: the required vector is, for example, $v = (x_2, -x_1, x_4, -x_3)$.

Case 2: two vectors p–orthonormal systems with support size 2.

If the p–orthonormal system S has two vectors with $|\text{supp}(S)| = 2$, the solution (valid for all $p \ge 1$) is as well trivial. Suppose, without loss of generality, that $\text{supp}(S) = \{1, 2\}$, $v_1 = (x_1, x_2, 0, 0)$ and $v_2 = (y_1, y_2, 0, 0)$. Then, the required vector is, for example, $v = (0, 0, x_1, x_2)$.

Case 3: three vectors p–orthonormal systems.

If the p–orthonormal system S has three vectors, their exterior product allows us to obtain the required vector (valid for all $p \ge 1$). It is enough to prove that all the coordinates of the exterior product are multiples of p and divide this vector by p to obtain the vector we are looking for.

So far, attempts to extend the proof of Problem 2 to arbitrary values of the natural number p have been unsuccessful, despite having been proven with a computer that the result is true up to $p = 10000$. This fact shows that the problem has a deep relationship with Number Theory. For discrete quantum computing the affirmative answer to Problem 1, as well as the proof of Conjectures 1 and 2, are very important. It would mean that discrete quantum computing maintains the most important properties relative to orthogonal and orthonormal vector systems and unitary transformations.

If we generalize Problem 2 by applying it to other dimensions, we see that counterexamples can be found for every dimension n that is not a multiple of 4. Thus, from Problem 2, we arrive at the following conjecture.

Conjecture 3. Given $n \equiv 0 \bmod 4$ $(n \ge 1)$ and $p \ge 1$ and a p–orthonormal system in $\mathbb{Z}^n$, S, then S can be extended to a p–orthonormal base.

In all the problems raised and the conjectures established, the parities of the coordinates are important and, where appropriate, their parity patterns. It is also interesting to note that if we only want orthogonal systems, without specifying the norm or level of the vector with which we want to extend the system, all problems and conjectures are solved affirmatively.

Finally, we want to comment that the authors of the work in which discrete quantum computing is related to Lagrange's four-square theorem [21], conjecture that Problem 1 has an affirmative answer.

DOI: http://dx.doi.org/10.5772/intechopen.98827

5. Does quantum physics allow discrete quantum computing?

Discrete quantum computing could in principle make error control easier. But in order to take advantage of the fact that quantum states are discrete, Quantum Physics must allow the construction of self-correcting systems. A system with these characteristics associates a basin of attraction with each discrete state so that whenever the $n-$qubit falls into said basin of attraction, the system automatically corrects it, transforming it into the associated discrete state. However, this process does not fulfill the Schrödinger equation because it is not unitary. And it cannot be the result of a quantum measurement either because the probability that the result was not the associated discrete state would not be zero. Then, how can Quantum Physics implement discrete quantum computing?

We believe that Quantum Physics can take one step further in the description of physical systems. Quantum Physics still fails to explain fundamental physical concepts, to the point that physicists as relevant as Feynman said "I think I can safely say that nobody understands quantum mechanics" and Quantum Mechanics has a reputation for being especially mysterious.

An example of a surprising result is the the no-cloning theorem [45–47], which states that it is impossible to create an independent and identical copy of an arbitrary unknown quantum state. This result of Quantum Physics contrasts with the self-reproducing systems of nature and is also derived from the Schrödinger equation, that predicts a unitary evolution of physical systems.

Quantum Physics has so far failed to explain the concepts for which it has acquired the fame of mysterious. We must assume that these mysteries are intrinsic to the nature of physical systems or that there is a road for Quantum Physics to explain them and open new paths for its development. Next we are going to analyze some of the less understandable concepts of Quantum Physics.

The first concept that is difficult to understand is the wave-particle duality. These concepts are inherently incompatible and nevertheless both are necessary to explain many results of Quantum Physics. If we assume that physical systems have a coherent physical description, we must conclude that elementary particles are neither waves nor particles. Therefore they must be something else.

On the other hand, the postulates of Quantum Physics introduce two processes to describe the evolution of physical systems: the Schrödinger equation and quantum measurements. The first predicts a unitary evolution of physical systems while the second seems to violate the prediction of the first. Many researchers assume that the result of the measurement of a quantum system is a random process whose probabilities depend on the measured system and not on the device that performs the measurement, and that the result is random, that is, there are no hidden variables that determine the result deterministically. In this interpretation the measurement process violates the Schrödinger equation. Other interpretations regard quantum states as statistical information about quantum systems, thus asserting that abrupt and discontinuous changes of quantum states are not problematic, simply reflecting updates of the available information. These reinforce the mysterious character of Quantum Physics and change its objective of describing physical systems for that of only obtaining information.

Finally, we want to comment on the interpretations made of the wave function obtained by solving the Schrodinger equation. It is common to hear that the wave function, for example of an electron, does not indicate that the particle is at all points where the wave function is not zero and that it is not an indicator of our ignorance of the position of the particle. On the one hand we give all the credit to the Schrödinger equation and on the other we take it away from the wave function.

As we see the controversy continues to haunt Quantum Physics. From our point of view, Quantum Physics has found a prediction system for the results of the measurements of physical systems, but it does not describe them. This prevents Quantum Physics from advancing in the deductive knowledge of physical systems, leaving only the advance based on experimentation. Does Quantum Physics really describe everything we can know about physical systems? We do not believe it.

What can be done to get out of this loop? We believe that we should focus on the initial problem: the wave-particle duality. As we have indicated before, this dilemma indicates that elementary particles are neither waves nor particles. Therefore the first objective is to determine its nature. To do this, we must look for questions that can be answered through the design of experiments and that shed light on the nature of elementary particles. In our opinion the first important question is the following: In how many points of space can an elementary particle be simultaneously?

Physics, in addition to the problems of Quantum Physics already mentioned, also has serious problems to combine two of its most notable theories: General Relativity and Quantum Physics. Undoubtedly, any theory that goes in the direction of discretizing space must also consider the discretization of time. In our study we only intend to contribute ideas so that Quantum Physics can overcome the controversies that it is not able to explain. We do not start from the hypothesis that Quantum Physics must be a discretized theory, but we believe that it must be a theory that allows self-correction and that this property must allow the implementation of a discrete quantum computation.

In Quantum Physics, different types of discretization have been proposed, in addition to the one presented in this article. Thus, in [48] a discretization of the quantum state space is proposed in order to explain Born's rule for probabilities. The proposal, despite being very similar to the one we have presented in this article, has very different objectives. In [48] it is used to try to explain two of the most important interpretations of Quantum Physics: Many Worlds and Copenhagen interpretation. In our case the objective is to define a discrete quantum computing model allowing effective control of quantum errors. And this objective leads us to pose an important question, aimed at explaining the wave-particle duality: In how many points of space can an elementary particle be simultaneously?

5.1 Hypothesis on the nature of elementary particles

Elementary particles cannot be in only one position in space because they cannot explain their behavior as waves. Then, in how many positions can they be simultaneously? The answer can be a finite number greater than one, a countable infinite number, or even an uncountable infinite number. Due to the principle of simplicity, we are inclined to take as a working hypothesis that the answer is a finite number greater than one.

And what does it mean for a particle to be simultaneously at various points in space? In our hypothesis the particle orbit between all its possible positions but being in only one at each time. Therefore simultaneity must be taken in a non-strict sense. That a particle orbits in different points means that it disappears from one point and appears in another and so on. The particle does not travel from one point to another through ordinary space and, in this sense, it may violate the special relativistic principle of speed limitation. Colloquially speaking the particle travels through a "wormhole", without deforming space through large concentrations of mass.

And, why do we choose this elementary particle model as a hypothesis? Because as we have said, the particle must be able to be in more than one point simultaneously and there are already experimental results of quantum nonlocality [49–53]. As far as we know, quantum nonlocality does not allow for faster-than-light communication and it is generally assumed that is compatible with special relativity and its universal speed limit of objects. We believe that quantum nonlocality in some sense violates the aforementioned principle of special relativity. We do not believe that the physical characteristics of the systems should be subordinated to the ability to transmit information.

From our point of view, the multi-position structure of the particles generates nonlocality in the usual space and breaks its Euclidean behavior. In this way physical systems can interact non-locally in space through their multi-position structure.

Another question that arises naturally from our working hypothesis is how scattered can the points that define an elementary particle be in space? Non-point particles can naturally explain their intrinsic angular momentum and this, in turn, give us information about the structure of the particles. For example, a particle that could be in three points in space would have an angular momentum proportional to the area of the triangle determined by its positions. This would indicate that the dispersion of the particles would occur on typical scales of Quantum Physics.

The multi-position particle hypothesis would again bring up some problems that originated Quantum Theories, such as, for example, the stability of atoms. This problem would be solved by the spatial scattering of the electrons around the nucleus. In this case the far electromagnetic field generated by the electrons would decrease faster than the inverse of the square of the distance and this would prevent the electrons from losing their energy in the form of electromagnetic radiation.

Our hypothesis would force us to readapt Quantum Theory. Therefore, we should plan experiments that allow us to contrast it. Is this possible?

5.2 How to test the hypothesis experimentally?

We would like to propose a couple of experiments that could theoretically provide information on our hypothesis about the structure of elementary particles. The first is a variation of the flagship experiment in which the wave-particle duality

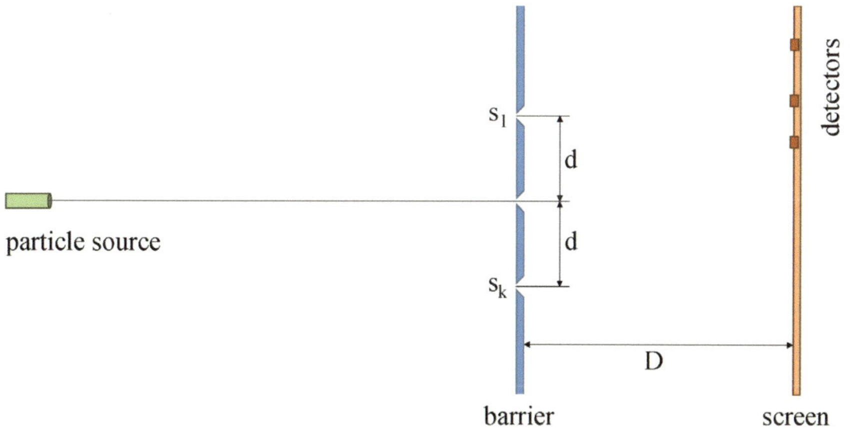

Figure 4.
k-slit experiment.

of elementary particles is tested: the double-slit experiment. The second uses a known quantum effect: the quantum tunneling.

Experiment 1. k-slits. We launch, one by one, elementary particles towards a barrier orthogonal to the direction of the movement of the particles (see **Figure 4**). In the barrier there are k parallel slits at a distance d one from the following: $s_1, s_2, \ldots, s_k$. Behind we place a screen parallel to the barrier and at a distance D from it. On this screen we place the detectors to obtain the interference pattern of the particles.

The objective of this experiment is to determine if the particles, according to our hypothesis, can be simultaneously in exactly $k-1$ positions. If this hypothesis is true, a particle cannot pass through the k slits. It can pass through $k-1$ slits at most. Therefore, the interference pattern will depend on whether the hypothesis is true.

We start the experiment by choosing $k=3$. If the hypothesis that the particles are in exactly $k-1$ positions simultaneously is not corroborated, we increase the value of k by 1 and carry out the experiment again. And when is our hypothesis confirmed? When the interference pattern obtained is $P(\text{true})$ instead of $P(\text{false})$:

1. $P(\text{true}) = \frac{P(s_1, \ldots, s_{k-1}) + P(s_1, \ldots, s_{k-2}, s_k) + \cdots + P(s_2, \ldots, s_k)}{k}$.

2. $P(\text{false}) = P(s_1, s_2, \ldots, s_k)$.

It would be necessary to estimate if the measurements can be precise enough to distinguish the two patterns and, in the first, if the probability of the k possible cases is the same or not.

Experiment 2. Quantum tunneling. We launch, one by one, elementary particles towards a potential barrier orthogonal to the direction of the movement of the particles (see **Figure 5(a)**). The energy of the particles is insufficient to jump the potential barrier and its width is small enough to allow the particles to have an appreciable probability of passing the barrier by tunneling. The particles are prepared in two different states. In the first state the intrinsic angular momentum of the particles is parallel to the direction of motion and, in the second state, it is orthogonal.

The objective of this experiment is to determine if the state of the particles influences the probability of quantum tunneling. If this influence is confirmed, it would mean that the orientation of the intrinsic angular momentum of the particles determines in some way the internal structure of the particle against the potential barrier. This could be explained quite understandably with the hypothesis that the particles are in exactly 3 positions at the same time. In this case the particle is always in a plane and the intrinsic angular momentum can orient that plane. If the three

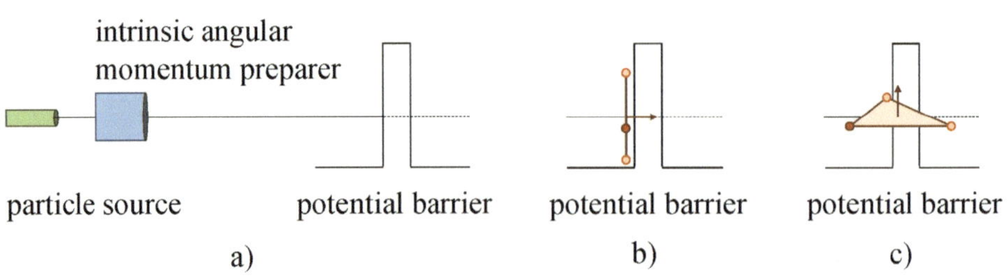

Figure 5.
Quantum tunneling experiment.

positions that define the particle reach the barrier simultaneously, the particle will not be able to pass (see **Figure 5(b)**). But if one of the positions arrives earlier, this position could cross the barrier while the particle orbits in the other positions (see **Figure 5(c)**). Thus, when the particle orbits in this position it will already be on the other side of the barrier.

We believe that it is not difficult to design more experiments that can shed light on our hypothesis of elementary particles. At this moment we are studying the dynamics of these multi-position particles.

6. Conclusions

In this article we introduce the discrete quantum computing as an alternative road to real quantum computing. The discrete quantum computing model is of great interest in itself both because, while maintaining all the important properties of quantum computing, it is an especially simplicity model and because error control is theoretically easier in this model. The introduced discrete quantum computing model satisfies some surprising properties that the authors believed would not hold and has deep connections to Number Theory.

The reason we set out on this alternative road to quantum computing is because error control in quantum computing is an extremely difficult challenge. The fact that the quantum computing model is continuous means that the golden rule of error control cannot be used: small errors are exactly corrected. A quantum computer is a very complex system from the point of view of error control. It allows reaching any quantum state (solution to the instance of a problem) by any path (algorithm). Doing this while keeping the error (entropy?) controlled is certainly an impressive challenge. As a consequence of the difficulty of controlling errors in continuous systems, there is no analog (continuous) device remotely comparable in operational complexity to a computer.

However, Quantum Physics does not allow the implementation of a discrete quantum computing model that allows self-correction of errors. To overcome this difficulty we ask Quantum Physics to go one step further in describing physical systems, beyond the prediction of measurement results. For this we propose a hypothesis about the nature of elementary particles that tries to overcome the never-understandable principle of wave-particle duality.

Summarizing, we propose an alternative road to quantum computing that passes through the discretization of this computing model and overcoming the interpretation gaps of Quantum Physics relative to the physical systems.

Author details

Jesús Lacalle
Universidad Politécnica de Madrid, Madrid, Spain

*Address all correspondence to: jesus.glopezdelacalle@upm.es

References

[1] Marella S T, Parisa H S K. Introduction to Quantum Computing. In: IntechOpen. DOI: 10.5772/intechopen.94103. Available from: https://www.intechopen.com/online-first/introduction-to-quantum-computing

[2] Nielsen M A, Chuang I L. Quantum Computation and Quantum Information. Cambridge University Press; 2010. 664 p. DOI: 10.1017/CBO9780511976667

[3] Calderbank A R, Shor P W. Good quantum error-correcting codes exist. Phys. Rev. A. 1996;54:1098–1105.

[4] Steane A M. Multiple particle inference and quantum error correction. Proc. Roy. Soc. A. 1996;452:2551.

[5] Gottesman D. Class of quantum error correcting codes saturating the quantum Hamming boud. Phys. Rev. A 1996;54: 1862.

[6] Calderbank A R, Rains E M, Shor P W, Sloane N J *A. quantum* Error Correction and Orthogonal Geometry. Phys. Rev. Lett. 1997;78:405.

[7] Gottesman D. Stabilizer Codes and Quantum Error Correction [thesis]. California Institute of Technology; 1997.

[8] Calderbank A R, Rains E M, Shor P W, Sloane N J *A. quantum* error correction via codes over GF(4). IEEE Trans. Inf. Theory. 1998;44(4):1369–1387.

[9] Shor P W. Fault-tolerant quantum computation. In: Proceedings of 37th Conference on Foundations of Computer Science; 14-16 October 1996; Burlington, VT, USA. IEEE Comput. Soc. Press; 1996. p. 56–65. arXiv:quant-ph/9605011. DOI: 10.1109/SFCS.1996.548464.

[10] Steane A M. Active stabilization, quantum computation and quantum state synthesis. Phys. Rev. Lett. 1997;78: 2252.

[11] Preskill J. Reliable quantum computers. Proc. Roy. Soc. Lond. A. 1998;454:385–410.

[12] Gottesman D. Theory of fault-tolerant quantum computation. Phys. Rev. A. 1998;57:127–137.

[13] Knill E, Laflamme R, Zurek W H. Resilient Quantum Computation. Science. 1998;279(5349):342–345. arXiv:quant-ph/9702058v1. DOI: 10.1126/science.279.5349.342

[14] Kitaev A Yu. Fault-tolerant quantum computation by anyons. Annals of Physics. 2003;303(1):2–30. arXiv:quant-ph/9707021. DOI: 10.1016/S0003-4916(02)00018-0

[15] Aharonov D, Ben-Or M. Fault-Tolerant Quantum Computation with Constant Error Rate. SIAM Journal on Computing. 2008;38(4):1207–1282. arXiv:quant-ph/9906129. DOI: 10.1137/S0097539799359385

[16] Lacalle, J., Pozo-Coronado, L.M., Fonseca de Oliveira, A.L., Quantum codes do not fix isotropic errors. Quantum Inf Process 20, 37 (2021). https://doi.org/10.1007/s11128-020-02980-3.

[17] Lacalle J, Pozo Coronado L M, Fonseca de Oliveira A L, Martín-Cuevas R. Quantum codes do not fix qubit independent errors. Will appear in American Journal of Information Science and Technology. 2021. arXiv: 2101.03971 [quant-ph]

[18] Lacalle J, Pozo Coronado L M. Variance of the sum of independent quantum computing errors. Quantum Information & Computation. 2019;19 (15-16):1294–1312. DOI: 10.26421/QIC19.15-16.

[19] Lacalle J, Pozo Coronado L M, Fonseca de Oliveira A L, Martín-Cuevas R. Quantum codes do not increase fidelity against isotropic errors. Personal communication 2021. It will appear in arXiv [quant-ph]

[20] Gatti L N, Lacalle J. A model of discrete quantum computation. Quantum Inf Process. 2018;17(192). DOI: 10.1007/s11128-018-1956-0

[21] Lacalle J, Gatti L N. Discrete quantum computation and Lagrange's four-square theorem. Quantum Inf Process. 2020;19(34). DOI: 10.1007/s11128-019-2528-7

[22] Gaitan F. Quantum error correction and fault tolerant quantum computing, CRC Press; 2008. 292 p.

[23] Bennet C H, DiVincenzo D P, Smolin J A, Wootters W K. Mixed state entanglement and quantum error correction. Los Alamos Physics Preprint Archive. 1999. arXiv:9909058 [quant-ph].

[24] Laflamme R, Miquel C, Paz J-P, Zurek W H. Perfect quantum error correction codes. Phys. Rev. Lett. 1996; 77:198. arXiv:9602019 [quant-ph].

[25] Benjamin S, Westmoreland M D. Modal quantum theory. Found. Phys. 2012;42(7):918--925.

[26] Ellerman D. Quantum mechanics over sets. arXiv:1310.8221v1 [quant-ph].

[27] Hanson A J, Ortiz G, Sabry A, Tai Y-T. Geometry of discrete quantum computing. J. Phys. A Math. Theor. 2013;46(18):185301.

[28] Hanson A J, Ortiz G, Sabry A, Tai Y-T. Discrete quantum theories. J. Phys. A Math. Theor. 2014;47(11):115305.

[29] Gatti L N, García-López J. Geometría de estados discretos en computación cuántica. In: 10th Andalusian Meeting on Discrete Mathematics; July 10-11, 2017; La Línea de la Concepción, Cádiz, Spain.

[30] Chandrashekar C M, Srikanth R, Laflamme R. Optimizing the discrete time quantum walk using a su(2) coin. Phys. Rev. A. 2008;77:032326.

[31] Lloyd S, Dreyer O. The universal path integral. Quant. Inf. Process. 2016; 15(2):959–967.

[32] Long G-L. General quantum interference principle and duality computer. Commun. Theor. Phys. 2006; 45(5):825.

[33] Long G-L, Liu Y. Duality computing in quantum computers. Commun. Theor. Phys. 2008;50(6):1303.

[34] Long G-L, Liu Y, Wang C. Allowable generalized quantum gates. Commun. Theor. Phys. 2009;51(1):65.

[35] Gudder S. Mathematical theory of duality quantum computers. Quant. Inf. Process. 2007;6(1):37–48.

[36] Long G-L. Mathematical theory of the duality computer in the density matrix formalism. Quant. Inf. Process. 2007;6(1):49–54.

[37] Wei S-J, Long G-L. Duality quantum computer and the efficient quantum simulations. Quant. Inf. Process. 2016;15 (3):1189–1212.

[38] Lomonaco S J. How to build a device that cannot be built. Quant. Inf. Process. 2016;15(3):1043–1056.

[39] Kitaev A Y, Shen A, Vyalyi M N. Classical and Quantum Computation. American Mathematical Society. Providence; 2002;47.

[40] Oscar B P, Mor T, Pulver M, Roychowdhury V, Vatan F. On universal and fault-tolerant quantum computing: a novel basis and a new

constructive proof of universality for shor's basis. In: 40th Annual Symposium on Foundations of Computer Science; October, 17-19,1999; New York City, NY, USA.

[41] Shi Y. Both toffoli and controlled-not need little help to do universal quantum computing. Quant. Inf. Comput. 2003;3(1):84–92.

[42] Aharonov D. A simple proof that Toffoli and Hadamard are quantum universal. arXiv:0301040] [quant-ph].

[43] Kliuchnikov V, Maslov D, Mosca M. Fast and efficient exact synthesis of single qubit unitaries generated by Clifford and T gates. Quant. Inf. Comput. 2013;13(7–8):607–630.

[44] Lagrange J L. Démonstration d'un théorème d'arithmétique, Oeuvres complétes 3; 1869. 189–201

[45] Park J. The concept of transition in quantum mechanics. Foundations of Physics. 1970;1(1):23–33. DOI: 10.1007/BF00708652

[46] Wootters W, Zurek W. A Single Quantum Cannot be Cloned. Nature. 1982;299(5886):802–803. DOI: 10.1038/299802a0

[47] Dieks D. Communication by EPR devices. Physics Letters A. 1982;92(6): 271–272. DOI: 10.1016/0375-9601(82)90084-6

[48] Buniy R V, Hsua S D H, Zee, A. Discreteness and the origin of probability in quantum mechanics. Physics Letters B. 2006;640: 219–223. DOI: 10.1016/j.physletb.2006.07.050

[49] Aspect A, Dalibard J, Roger G. Experimental Test of Bell's Inequalities Using Time- Varying Analyzers. Physical Review Letters. 1982;49(25): 1804–1807. DOI: 10.1103/PhysRevLett.49.1804

[50] Rowe M, Kielpinski D, Meyer V et al. Experimental violation of a Bell's Inequality with efficient detection. Nature. 2001;409(6822):791–794. DOI: 10.1038/35057215

[51] Hensen B, Bernien H, Dréau A et al. Loophole-free Bell inequality violation using electron spins separated by 1.3 kilometres. Nature. 2015;526(7575):682–686. DOI: 10.1038/nature15759

[52] Giustina M, Versteegh M A M et al. Significant-Loophole-Free Test of Bell's Theorem with Entangled Photons. Physical Review Letters. 2015;115(25): 250401. DOI: 10.1103/PhysRevLett.115.250401

[53] Shalm L K, Meyer-Scott E et al. (December 2015). Strong Loophole-Free Test of Local Realism. Physical Review Letters. 2015;115(25):250402. DOI: 10.1103/PhysRevLett.115.250402.

Chapter 6

Obscure Qubits and Membership Amplitudes

Steven Duplij and Raimund Vogl

Abstract

We propose a concept of quantum computing which incorporates an additional kind of uncertainty, i.e. vagueness (fuzziness), in a natural way by introducing new entities, obscure qudits (e.g. obscure qubits), which are characterized simultaneously by a quantum probability and by a membership function. To achieve this, a membership amplitude for quantum states is introduced alongside the quantum amplitude. The Born rule is used for the quantum probability only, while the membership function can be computed from the membership amplitudes according to a chosen model. Two different versions of this approach are given here: the "product" obscure qubit, where the resulting amplitude is a product of the quantum amplitude and the membership amplitude, and the "Kronecker" obscure qubit, where quantum and vagueness computations are to be performed independently (i.e. quantum computation alongside truth evaluation). The latter is called a double obscure-quantum computation. In this case, the measurement becomes mixed in the quantum and obscure amplitudes, while the density matrix is not idempotent. The obscure-quantum gates act not in the tensor product of spaces, but in the direct product of quantum Hilbert space and so called membership space which are of different natures and properties. The concept of double (obscure-quantum) entanglement is introduced, and vector and scalar concurrences are proposed, with some examples being given.

Keywords: qubit, fuzzy, membership function, amplitude, Hilbert space

1. Introduction

Nowadays, the development of quantum computing technique is governed by theoretical extensions of its ground concepts [1–3]. One of them is to allow two kinds of uncertainty, sometimes called randomness and vagueness/fuzziness (for a review, see, [4]), which leads to the formulation of combined probability and possibility theories [5] (see, also, [6–9]). Various interconnections between vagueness and quantum probability calculus were considered in [10–13], including the treatment of inaccuracy in measurements [14, 15], non-sharp amplitude densities [16] and the related concept of partial Hilbert spaces [17].

Relations between truth values and probabilities were also given in [18]. The hardware realization of computations with vagueness was considered in [19, 20]. On the fundamental physics side, it was shown that the discretization of space–time at small distances can lead to a discrete (or fuzzy) character for the quantum states themselves [21–24].

With a view to applications of the above ideas in quantum computing, we introduce a definition of quantum state which is described by both a quantum probability and a membership function, and thereby incorporate vagueness/fuzziness directly into the formalism. In addition to the probability amplitude we will define a membership amplitude, and such a state will be called an obscure/fuzzy qubit (or qudit).

In general, the Born rule will apply to the quantum probability alone, while the membership function can be taken to be an arbitrary function of all the amplitudes fixed by the chosen model of vagueness. Two different models of "obscure-quantum computations with truth" are proposed below: (1) A "Product" obscure qubit, in which the resulting amplitude is the product (in $\mathbb{C}$) of the quantum amplitude and the membership amplitude; (2) A "Kronecker" obscure qubit for which computations are performed "in parallel", so that quantum amplitudes and the membership amplitudes form "vectors", which we will call obscure-quantum amplitudes. In the latter case, which we call a double obscure-quantum computation, the protocol of measurement depends on both the quantum and obscure amplitudes, and in this case the density matrix need not be idempotent. We define a new kind of "gate", namely, the obscure-quantum gates, which are linear transformations in the direct product (not in the tensor product) of spaces: a quantum Hilbert space and a so-called membership space having special fuzzy properties. We introduce a new concept of double (obscure-quantum) entanglement, in which vector and scalar concurrences are defined and computed for some examples.

2. Preliminaries

To establish notation standard in the literature (see, e.g. [1, 2, 25–27]) we present the following definitions. In an underlying d-dimensional Hilbert space, the standard qudit (using the computational basis and Dirac notation) $\mathcal{H}_q^{(d)}$ is given by

$$\left|\psi^{(d)}\right\rangle = \sum_{i=0}^{d-1} a_i |i\rangle, \qquad a_i \in \mathbb{C}, |i\rangle \in \mathcal{H}_q^{(d)}, \tag{1}$$

where a_i is a probability amplitude of the state $|i\rangle$. (For a review, see, e.g. [28, 29]) The probability p_i to measure the ith state is $p_i = F_{p_i}(a_1, \dots, a_n)$, $0 \leq p_i \leq 1$, $0 \leq i \leq d-1$. The shape of the functions F_{p_i} is governed by the Born rule $F_{p_i}(a_1, \dots, a_d) = |a_i|^2$, and $\sum_{i=0}^{d} p_i = 1$. A one-qudit ($L = 1$) quantum gate is a unitary transformation $U^{(d)} : \mathcal{H}_q^{(d)} \to \mathcal{H}_q^{(d)}$ described by unitary $d \times d$ complex matrices acting on the vector (1), and for a register containing L qudits quantum gates are unitary $d^L \times d^L$ matrices. The quantum circuit model [30, 31] forms the basis for the standard concept of quantum computing. Here the quantum algorithms are compiled as a sequence of elementary gates acting on a register containing L qubits (or qudits), followed by a measurement to yield the result [25, 32].

For further details on qudits and their transformations, see for example the reviews [28, 29] and the references therein.

3. Membership amplitudes

We define an obscure qudit with d states via the following superposition (in place of that given in (1))

$$\left|\psi_{ob}^{(d)}\right\rangle = \sum_{i=1}^{d-1} \alpha_i a_i |i\rangle, \qquad (2)$$

where a_i is a (complex) probability amplitude $a_i \in \mathbb{C}$, and we have introduced a (real) membership amplitude α_i, with $\alpha_i \in [0,1]$, $0 \leq i \leq d-1$. The probability p_i to find the ith state upon measurement, and the membership function μ_i ("of truth") for the ith state are both functions of the corresponding amplitudes as follows

$$p_i = F_{p_i}(a_0, \dots, a_{d-1}), \qquad 0 \leq p_i \leq 1, \qquad (3)$$

$$\mu_i = F_{\mu_i}(\alpha_0, \dots, \alpha_{d-1}), \qquad 0 \leq \mu_i \leq 1. \qquad (4)$$

The dependence of the probabilities of the ith states upon the amplitudes, i.e. the form of the function F_{p_i} is fixed by the Born rule

$$F_{p_i}(a_1, \dots, a_n) = |a_i|^2, \qquad (5)$$

while the form of F_{μ_i} will vary according to different obscurity assumptions. In this paper we consider only real membership amplitudes and membership functions (complex obscure sets and numbers were considered in [33–35]). In this context the real functions F_{p_i} and F_{μ_i}, $0 \leq i \leq d-1$ will contain complete information about the obscure qudit (2).

We impose the normalization conditions

$$\sum_{i=0}^{d-1} p_i = 1, \qquad (6)$$

$$\sum_{i=0}^{d-1} \mu_i = 1, \qquad (7)$$

where the first condition is standard in quantum mechanics, while the second condition is taken to hold by analogy. Although (7) may not be satisfied, we will not consider that case.

For $d = 2$, we obtain for the obscure qubit the general form (instead of that in (2))

$$\left|\psi_{ob}^{(2)}\right\rangle = \alpha_0 a_0 |0\rangle + \alpha_1 a_1 |1\rangle, \qquad (8)$$

$$F_{p_0}(a_0, a_1) + F_{p_1}(a_0, a_1) = 1, \qquad (9)$$

$$F_{\mu_0}(\alpha_0, \alpha_1) + F_{\mu_1}(\alpha_0, \alpha_1) = 1. \qquad (10)$$

The Born probabilities to observe the states $|0\rangle$ and $|1\rangle$ are

$$p_0 = F_{p_0}^{Born}(a_0, a_1) = |a_0|^2, \qquad p_1 = F_{p_1}^{Born}(a_0, a_1) = |a_1|^2, \qquad (11)$$

and the membership functions are

$$\mu_0 = F_{\mu_0}(\alpha_0, \alpha_1), \qquad \mu_1 = F_{\mu_1}(\alpha_0, \alpha_1). \qquad (12)$$

If we assume the Born rule (11) for the membership functions as well

$$F_{\mu_0}(\alpha_0, \alpha_1) = \alpha_0^2, \qquad F_{\mu_1}(\alpha_0, \alpha_1) = \alpha_1^2, \qquad (13)$$

(which is one of various possibilities depending on the chosen model), then

$$|a_0|^2 + |a_1|^2 = 1, \tag{14}$$

$$\alpha_0^2 + \alpha_1^2 = 1. \tag{15}$$

Using (14)–(15) we can parametrize (8) as

$$\left|\psi_{ob}^{(2)}\right\rangle = \cos\frac{\theta}{2}\cos\frac{\theta_\mu}{2}|0\rangle + e^{i\varphi}\sin\frac{\theta}{2}\sin\frac{\theta_\mu}{2}|1\rangle, \tag{16}$$

$$0 \le \theta \le \pi, \qquad 0 \le \varphi \le 2\pi, \qquad 0 \le \theta_\mu \le \pi. \tag{17}$$

Therefore, obscure qubits (with Born-like rule for the membership functions) are geometrically described by a pair of vectors, each inside a Bloch ball (and not as vectors on the boundary spheres, because "$|\sin|, |\cos| \le 1$"), where one is for the probability amplitude (an ellipsoid inside the Bloch ball with $\theta_\mu = const_1$), and the other for the membership amplitude (which is reduced to an ellipse, being a slice inside the Bloch ball with $\theta = const_2$, $\varphi = const_3$). The norm of the obscure qubits is not constant however, because

$$\left\langle\psi_{ob}^{(2)}|\psi_{ob}^{(2)}\right\rangle = \frac{1}{2} + \frac{1}{4}\cos\left(\theta + \theta_\mu\right) + \frac{1}{4}\cos\left(\theta - \theta_\mu\right). \tag{18}$$

In the case where $\theta = \theta_\mu$, the norm (18) becomes $1 - \frac{1}{2}\sin^2\theta$, reaching its minimum $\frac{1}{2}$ when $\theta = \theta_\mu = \frac{\pi}{2}$.

Note that for complicated functions $F_{\mu_{0,1}}(\alpha_0, \alpha_1)$ the condition (15) may be not satisfied, but the condition (7) should nevertheless always be valid. The concrete form of the functions $F_{\mu_{0,1}}(\alpha_0, \alpha_1)$ depends upon the chosen model. In the simplest case, we can identify two arcs on the Bloch ellipse for α_0, α_1 with the membership functions and obtain

$$F_{\mu_0}(\alpha_0, \alpha_1) = \frac{2}{\pi}\arctan\frac{\alpha_1}{\alpha_0}, \tag{19}$$

$$F_{\mu_1}(\alpha_0, \alpha_1) = \frac{2}{\pi}\arctan\frac{\alpha_0}{\alpha_1}, \tag{20}$$

such that $\mu_0 + \mu_1 = 1$, as in (7).

In [36, 37] a two stage special construction of quantum obscure/fuzzy sets was considered. The so-called classical-quantum obscure/fuzzy registers were introduced in the first step (for $n = 2$, the minimal case) as

$$|s\rangle_f = \sqrt{1-f}|0\rangle + \sqrt{f}|1\rangle, \tag{21}$$

$$|s\rangle_g = \sqrt{1-g}|0\rangle + \sqrt{g}|1\rangle, \tag{22}$$

where $f, g \in [0, 1]$ are the relevant classical-quantum membership functions. In the second step their quantum superposition is defined by

$$|s\rangle = c_f|s\rangle_f + c_g|s\rangle_g, \tag{23}$$

where c_f and c_g are the probability amplitudes of the fuzzy states $|s\rangle_f$ and $|s\rangle_g$, respectively. It can be seen that the state (23) is a particular case of (8) with

$$\alpha_0 a_0 = c_f \sqrt{1-f} + c_g \sqrt{1-g}, \tag{24}$$

$$\alpha_1 a_1 = c_f \sqrt{f} + c_g \sqrt{g}. \tag{25}$$

This gives explicit connection of our double amplitude description of obscure qubits with the approach [36, 37] which uses probability amplitudes and the membership functions. It is important to note that the use of the membership amplitudes introduced here α_i and (2) allows us to exploit the standard quantum gates, but not to define new special ones, as in [36, 37].

Another possible form of $F_{\mu_{0,1}}(\alpha_0, \alpha_1)$ (12), with the corresponding membership functions satisfying the standard fuzziness rules, can be found using a standard homeomorphism between the circle and the square. In [38, 39] this transformation was applied to the probability amplitudes $a_{0,1}$, but here we exploit it for the membership amplitudes $\alpha_{0,1}$

$$F_{\mu_0}(\alpha_0, \alpha_1) = \frac{2}{\pi} \arcsin \sqrt{\frac{\alpha_0^2 * \operatorname{sign} \alpha_0 - \alpha_1^2 * \operatorname{sign} \alpha_1 + 1}{2}}, \tag{26}$$

$$F_{\mu_1}(\alpha_0, \alpha_1) = \frac{2}{\pi} \arcsin \sqrt{\frac{\alpha_0^2 * \operatorname{sign} \alpha_0 + \alpha_1^2 * \operatorname{sign} \alpha_1 + 1}{2}}. \tag{27}$$

So for positive $\alpha_{0,1}$ we obtain (cf. [38])

$$F_{\mu_0}(\alpha_0, \alpha_1) = \frac{2}{\pi} \arcsin \sqrt{\frac{\alpha_0^2 - \alpha_1^2 + 1}{2}}, \tag{28}$$

$$F_{\mu_1}(\alpha_0, \alpha_1) = 1. \tag{29}$$

The equivalent membership functions for the outcome are

$$\max\left(\min\left(F_{\mu_0}(\alpha_0, \alpha_1), 1 - F_{\mu_1}(\alpha_0, \alpha_1)\right), \min\left(1 - F_{\mu_0}(\alpha_0, \alpha_1)\right), F_{\mu_1}(\alpha_0, \alpha_1)\right), \tag{30}$$

$$\min\left(\max\left(F_{\mu_0}(\alpha_0, \alpha_1), 1 - F_{\mu_1}(\alpha_0, \alpha_1)\right), \max\left(1 - F_{\mu_0}(\alpha_0, \alpha_1)\right), F_{\mu_1}(\alpha_0, \alpha_1)\right). \tag{31}$$

There are many different models for $F_{\mu_{0,1}}(\alpha_0, \alpha_1)$ which can be introduced in such a way that they satisfy the obscure set axioms [7, 9].

4. Transformations of obscure qubits

Let us consider the obscure qubits in the vector representation, such that

$$|0\rangle = \begin{pmatrix} 1 \\ 0 \end{pmatrix}, \qquad |1\rangle = \begin{pmatrix} 0 \\ 1 \end{pmatrix} \tag{32}$$

are basis vectors of $\mathcal{H}_q^{(2)}$. Then a standard quantum computational process in the quantum register with L obscure qubits (qudits (1)) is performed by sequences of unitary matrices $\hat{U}$ of size $2^L \times 2^L$ $(n^L \times n^L)$, $\hat{U}^\dagger \hat{U} = \hat{I}$, which are called quantum gates ($\hat{I}$ is the unit matrix). Thus, for one obscure qubit the quantum gates are 2×2 unitary complex matrices.

In the vector representation, an obscure qubit differs from the standard qubit (8) by a 2×2 invertible diagonal (not necessarily unitary) matrix

$$\left|\psi_{ob}^{(2)}\right\rangle = \hat{M}(\alpha_0, \alpha_1)\left|\psi^{(2)}\right\rangle, \tag{33}$$

$$\hat{M}(\alpha_0, \alpha_1) = \begin{pmatrix} \alpha_0 & 0 \\ 0 & \alpha_1 \end{pmatrix}. \tag{34}$$

We call $\hat{M}(\alpha_0, \alpha_1)$ a membership matrix which can optionally have the property

$$\mathrm{tr}\hat{M}^2 = 1, \tag{35}$$

if (15) holds.

Let us introduce the orthogonal commuting projection operators

$$\hat{P}_0 = \begin{pmatrix} 1 & 0 \\ 0 & 0 \end{pmatrix}, \qquad \hat{P}_1 = \begin{pmatrix} 0 & 0 \\ 0 & 1 \end{pmatrix}, \tag{36}$$

$$\hat{P}_0^2 = \hat{P}_0, \quad \hat{P}_1^2 = \hat{P}_1, \quad \hat{P}_0\hat{P}_1 = \hat{P}_1\hat{P}_0 = \hat{0}, \tag{37}$$

where $\hat{0}$ is the 2×2 zero matrix. Well-known properties of the projections are that

$$\hat{P}_0\left|\psi^{(2)}\right\rangle = a_0|0\rangle, \qquad \hat{P}_1\left|\psi^{(2)}\right\rangle = a_1|0\rangle, \tag{38}$$

$$\left\langle\psi^{(2)}\middle|\hat{P}_0\middle|\psi^{(2)}\right\rangle = |a_0|^2, \qquad \left\langle\psi^{(2)}\middle|\hat{P}_1\middle|\psi^{(2)}\right\rangle = |a_1|^2. \tag{39}$$

Therefore, the membership matrix (34) can be defined as a linear combination of the projection operators with the membership amplitudes as coefficients

$$\hat{M}(\alpha_0, \alpha_1) = \alpha_0\hat{P}_0 + \alpha_1\hat{P}_1. \tag{40}$$

We compute

$$\hat{M}(\alpha_0, \alpha_1)\left|\psi_{ob}^{(2)}\right\rangle = \alpha_0^2 a_0|0\rangle + \alpha_1^2 a_1|1\rangle. \tag{41}$$

We can therefore treat the application of the membership matrix (33) as providing the origin of a reversible but non-unitary "obscure measurement" on the standard qubit to obtain an obscure qubit (cf. the "mirror measurement" [40, 41] and also the origin of ordinary qubit states on the fuzzy sphere [42]).

An obscure analog of the density operator (for a pure state) is the following form for the density matrix in the vector representation

$$\rho_{ob}^{(2)} = \left|\psi_{ob}^{(2)}\right\rangle\left\langle\psi_{ob}^{(2)}\right| = \begin{pmatrix} \alpha_0^2|a_0|^2 & \alpha_0 a_0^* \alpha_1 a_1 \\ \alpha_0 a_0 \alpha_1 a_1^* & \alpha_1^2|a_1|^2 \end{pmatrix} \tag{42}$$

with the obvious standard singularity property $\det \rho_{ob}^{(2)} = 0$. But $\mathrm{tr}\rho_{ob}^{(2)} = \alpha_0^2|a_0|^2 + \alpha_1^2|a_1|^2 \neq 1$, and here there is no idempotence $\left(\rho_{ob}^{(2)}\right)^2 \neq \rho_{ob}^{(2)}$, which distincts $\rho_{ob}^{(2)}$ from the standard density operator.

5. Kronecker obscure qubits

We next introduce an analog of quantum superposition for membership amplitudes, called "obscure superposition" (cf. [43], and also [44]).

Quantum amplitudes and membership amplitudes will here be considered separately in order to define an "obscure qubit" taking the form of a "double superposition" (cf. (8), and a generalized analog for qudits (1) is straightforward)

$$|\boldsymbol{\Psi}_{ob}\rangle = \frac{\hat{A}_0|\hat{0}\rangle + \hat{A}_1|\hat{1}\rangle}{\sqrt{2}}, \tag{43}$$

where the two-dimensional "vectors"

$$\hat{A}_{0,1} = \begin{bmatrix} a_{0,1} \\ \alpha_{0,1} \end{bmatrix} \tag{44}$$

are the (double) "obscure-quantum amplitudes" of the generalized states $|\hat{0}\rangle$, $|\hat{1}\rangle$. For the conjugate of an obscure qubit we put (informally)

$$\langle\boldsymbol{\Psi}_{ob}| = \frac{\hat{A}_0^{\star}\langle\hat{0}| + \hat{A}_1^{\star}\langle\hat{1}|}{\sqrt{2}}, \tag{45}$$

where we denote $\hat{A}_{0,1}^{\star} = \begin{bmatrix} a_{0,1}^{*} & \alpha_{0,1} \end{bmatrix}$, such that $\hat{A}_{0,1}^{\star}\hat{A}_{0,1} = |a_{0,1}|^2 + \alpha_{0,1}^2$. The (double) obscure qubit is "normalized" in such a way that, if the conditions (14)–(15) hold, then

$$\langle\boldsymbol{\Psi}_{ob}|\boldsymbol{\Psi}_{ob}\rangle = \frac{|a_0|^2 + |a_1|^2}{2} + \frac{\alpha_0^2 + \alpha_1^2}{2} = 1. \tag{46}$$

A measurement should be made separately and independently in the "probability space" and the "membership space" which can be represented by using an analog of the Kronecker product. Indeed, in the vector representation (32) for the quantum states and for the direct product amplitudes (44) we should have

$$|\boldsymbol{\Psi}_{ob}\rangle_{(0)} = \frac{1}{\sqrt{2}}\hat{A}_0 \otimes_K \begin{pmatrix} 1 \\ 0 \end{pmatrix} + \hat{A}_1 \otimes_K \begin{pmatrix} 0 \\ 1 \end{pmatrix}, \tag{47}$$

where the (left) Kronecker product is defined by (see (32))

$$\begin{bmatrix} a \\ \alpha \end{bmatrix} \otimes_K \begin{pmatrix} c \\ d \end{pmatrix} = \begin{bmatrix} a\begin{pmatrix} c \\ d \end{pmatrix} \\ \alpha\begin{pmatrix} c \\ d \end{pmatrix} \end{bmatrix} = \begin{bmatrix} a(c\hat{e}_0 + d\hat{e}_1) \\ \alpha(c\hat{e}_0 + d\hat{e}_1) \end{bmatrix}, \tag{48}$$
$$\hat{e}_0 = \begin{pmatrix} 1 \\ 0 \end{pmatrix}, \quad \hat{e}_1 = \begin{pmatrix} 0 \\ 1 \end{pmatrix}, \quad \hat{e}_{0,1} \in \mathcal{H}_q^{(2)}.$$

Informally, the wave function of the obscure qubit, in the vector representation, now "lives" in the four-dimensional space of (48) which has two two-dimensional spaces as blocks. The upper block, the quantum subspace, is the ordinary Hilbert space $\mathcal{H}_q^{(2)}$, but the lower block should have special (fuzzy) properties, if it is treated as an obscure (membership) subspace $\mathcal{V}^{(2)}_{memb}$. Thus, the four-dimensional space, where "lives" $\left|\Psi^{(2)}_{ob}\right\rangle$, is not an ordinary tensor product of vector spaces, because of (48), and the "vector" $\hat{A}$ on the l.h.s. has entries of different natures, that is the quantum amplitudes $a_{0,1}$ and the membership amplitudes $\alpha_{0,1}$. Despite the unit vectors in $\mathcal{H}^{(2)}_q$ and $\mathcal{V}^{(2)}_{memb}$ having the same form (32), they belong to different spaces (as they are vector spaces over different fields). Therefore, instead of (48) we introduce a "Kronecker-like product" $\tilde{\otimes}_K$ by

$$\begin{bmatrix} a \\ \alpha \end{bmatrix} \tilde{\otimes}_K \begin{pmatrix} c \\ d \end{pmatrix} = \begin{bmatrix} a(c\hat{e}_0 + d\hat{e}_1) \\ \alpha(c\varepsilon_0 + d\varepsilon_1) \end{bmatrix}, \tag{49}$$

$$\hat{e}_0 = \begin{pmatrix} 1 \\ 0 \end{pmatrix}, \quad \hat{e}_1 = \begin{pmatrix} 0 \\ 1 \end{pmatrix}, \quad \hat{e}_{0,1} \in \mathcal{H}^{(2)}_q, \tag{50}$$

$$\varepsilon_0 = \begin{pmatrix} 1 \\ 0 \end{pmatrix}^{(\mu)}, \quad \varepsilon_1 = \begin{pmatrix} 0 \\ 1 \end{pmatrix}^{(\mu)}, \quad \varepsilon_{0,1} \in \mathcal{V}^{(2)}_{memb}. \tag{51}$$

In this way, the obscure qubit (43) can be presented in the from

$$\begin{aligned} |\boldsymbol{\Psi}_{ob}\rangle &= \frac{1}{\sqrt{2}} \begin{bmatrix} a_0 \begin{pmatrix} 1 \\ 0 \end{pmatrix} \\ \alpha_0 \begin{pmatrix} 1 \\ 0 \end{pmatrix}^{(\mu)} \end{bmatrix} + \frac{1}{\sqrt{2}} \begin{bmatrix} a_1 \begin{pmatrix} 0 \\ 1 \end{pmatrix} \\ \alpha_1 \begin{pmatrix} 0 \\ 1 \end{pmatrix}^{(\mu)} \end{bmatrix} \\ &= \frac{1}{\sqrt{2}} \begin{bmatrix} a_0 \hat{e}_0 \\ \alpha_0 \varepsilon_0 \end{bmatrix} + \frac{1}{\sqrt{2}} \begin{bmatrix} a_1 \hat{e}_1 \\ \alpha_1 \varepsilon_1 \end{bmatrix}. \end{aligned} \tag{52}$$

Therefore, we call the double obscure qubit (52) a "Kronecker obscure qubit" to distinguish it from the obscure qubit (8). It can be also presented using the Hadamard product (the element-wise or Schur product)

$$\begin{bmatrix} a \\ \alpha \end{bmatrix} \otimes_H \begin{pmatrix} c \\ d \end{pmatrix} = \begin{bmatrix} ac \\ \alpha d \end{bmatrix} \tag{53}$$

in the following form

$$|\boldsymbol{\Psi}_{ob}\rangle = \frac{1}{\sqrt{2}} \hat{A}_0 \otimes_H \hat{E}_0 + \frac{1}{\sqrt{2}} \hat{A}_1 \otimes_H \hat{E}_1, \tag{54}$$

where the unit vectors of the total four-dimensional space are

$$\hat{E}_{0,1} = \begin{bmatrix} \hat{e}_{0,1} \\ \varepsilon_{0,1} \end{bmatrix} \in \mathcal{H}^{(2)}_q \times \mathcal{V}^{(2)}_{memb}. \tag{55}$$

The probabilities $p_{0,1}$ and membership functions $\mu_{0,1}$ of the states $|\hat{0}\rangle$ and $|\hat{1}\rangle$ are computed through the corresponding amplitudes by (11) and (12)

$$p_i = |a_i|^2, \qquad \mu_i = F_{\mu_i}(\alpha_0, \alpha_1), \qquad i = 0, 1, \tag{56}$$

and in the particular case by (13) satisfying (15).

By way of example, consider a Kronecker obscure qubit (with a real quantum part) with probability p and membership function μ (measure of "trust") of the state $|\hat{0}\rangle$, and of the state $|\hat{1}\rangle$ given by $1-p$ and $1-\mu$ respectively. In the model (19)–(20) for μ_i (which is not Born-like) we obtain

$$|\Psi_{ob}\rangle = \frac{1}{\sqrt{2}}\begin{bmatrix} \begin{pmatrix} \sqrt{p} \\ 0 \end{pmatrix} \\ \begin{pmatrix} \cos\frac{\pi}{2}\mu \\ 0 \end{pmatrix}^{(\mu)} \end{bmatrix} + \frac{1}{\sqrt{2}}\begin{bmatrix} \begin{pmatrix} 0 \\ \sqrt{1-p} \end{pmatrix} \\ \begin{pmatrix} 0 \\ \sin\frac{\pi}{2}\mu \end{pmatrix}^{(\mu)} \end{bmatrix} \tag{57}$$
$$= \frac{1}{\sqrt{2}}\begin{bmatrix} \hat{e}_0\sqrt{p} \\ \varepsilon_0\cos\frac{\pi}{2}\mu \end{bmatrix} + \frac{1}{\sqrt{2}}\begin{bmatrix} \hat{e}_1\sqrt{1-p} \\ \varepsilon_1\sin\frac{\pi}{2}\mu \end{bmatrix},$$

where $\hat{e}_i$ and ε_i are unit vectors defined in (50) and (51).

This can be compared e.g. with the "classical-quantum" approach (23) and [36, 37], in which the elements of columns are multiplied, while we consider them independently and separately.

6. Obscure-quantum measurement

Let us consider the case of one Kronecker obscure qubit register $L = 1$ (see (47)), or using (48) in the vector representation (52). The standard (double) orthogonal commuting projection operators, "Kronecker projections" are (cf. (36))

$$\mathbf{P}_0 = \begin{bmatrix} \hat{P}_0 & \hat{0} \\ \ddot{0} & \ddot{P}_0^{(\mu)} \end{bmatrix}, \qquad \mathbf{P}_1 = \begin{bmatrix} \hat{P}_1 & \hat{0} \\ \ddot{0} & \ddot{P}_1^{(\mu)} \end{bmatrix}, \tag{58}$$

where $\hat{0}$ is the 2×2 zero matrix, and $\hat{P}_{0,1}^{(\mu)}$ are the projections in the membership subspace $\mathcal{V}_{memb}^{(2)}$ (of the same form as the ordinary quantum projections $\hat{P}_{0,1}$ (36))

$$\hat{P}_0^{(\mu)} = \begin{pmatrix} 1 & 0 \\ 0 & 0 \end{pmatrix}^{(\mu)}, \qquad \hat{P}_1^{(\mu)} = \begin{pmatrix} 0 & 0 \\ 0 & 1 \end{pmatrix}^{(\mu)}, \qquad \hat{P}_0^{(\mu)}, \hat{P}_1^{(\mu)} \in \operatorname{End}\mathcal{V}_{memb}^{(2)}, \tag{59}$$

$$\hat{P}_0^{(\mu)2} = \hat{P}_0^{(\mu)}, \quad \hat{P}_1^{(\mu)2} = \hat{P}_1^{(\mu)}, \quad \hat{P}_0^{(\mu)}\hat{P}_1^{(\mu)} = \hat{P}_1^{(\mu)}\hat{P}_0^{(\mu)} = \hat{0}. \tag{60}$$

For the double projections we have (cf. (37))

$$\boldsymbol{P}_0^2 = \boldsymbol{P}_0, \;\; \boldsymbol{P}_1^2 = \boldsymbol{P}_1, \;\; \boldsymbol{P}_0\boldsymbol{P}_1 = \boldsymbol{P}_1\boldsymbol{P}_0 = \mathbf{0}, \tag{61}$$

where $\mathbf{0}$ is the 4×4 zero matrix, and $\boldsymbol{P}_{0,1}$ act on the Kronecker qubit (58) in the standard way (cf. (38))

$$\mathbf{P}_0|\Psi_{ob}\rangle = \frac{1}{\sqrt{2}}\begin{bmatrix} a_0\begin{pmatrix} 1 \\ 0 \end{pmatrix} \\ \alpha_0\begin{pmatrix} 1 \\ 0 \end{pmatrix}^{(\mu)} \end{bmatrix} = \frac{1}{\sqrt{2}}\begin{bmatrix} a_0\hat{e}_0 \\ \alpha_0\varepsilon_0 \end{bmatrix} = \frac{1}{\sqrt{2}}\hat{A}_0 \otimes_H \hat{E}_0, \tag{62}$$

$$\mathbf{P}_1|\boldsymbol{\Psi}_{ob}\rangle = \frac{1}{\sqrt{2}}\begin{bmatrix} a_1\begin{pmatrix}0\\1\end{pmatrix} \\ \alpha_1\begin{pmatrix}0\\1\end{pmatrix}^{(\mu)} \end{bmatrix} = \frac{1}{\sqrt{2}}\begin{bmatrix} a_1\hat{e}_1 \\ \alpha_1\varepsilon_1 \end{bmatrix} = \frac{1}{\sqrt{2}}\hat{A}_1 \otimes_H \hat{E}_1. \tag{63}$$

Observe that for Kronecker qubits there exist in addition to (58) the following orthogonal commuting projection operators

$$\mathbf{P}_{01} = \begin{bmatrix} \hat{P}_0 & \hat{0} \\ \hat{0} & \hat{P}_1^{(\mu)} \end{bmatrix}, \qquad \mathbf{P}_{10} = \begin{bmatrix} \hat{P}_1 & \hat{0} \\ \hat{0} & \hat{P}_0^{(\mu)} \end{bmatrix}, \tag{64}$$

and we call these the "crossed" double projections. They satisfy the same relations as (61)

$$\boldsymbol{P}_{01}^2 = \boldsymbol{P}_{01}, \ \boldsymbol{P}_{10}^2 = \boldsymbol{P}_{10}, \ \boldsymbol{P}_{01}\boldsymbol{P}_{10} = \boldsymbol{P}_{10}\boldsymbol{P}_{01} = \mathbf{0}, \tag{65}$$

but act on the obscure qubit in a different ("mixing") way than (62) i.e.

$$\mathbf{P}_{01}|\boldsymbol{\Psi}_{ob}\rangle = \frac{1}{\sqrt{2}}\begin{bmatrix} a_0\begin{pmatrix}1\\0\end{pmatrix} \\ \alpha_1\begin{pmatrix}0\\1\end{pmatrix} \end{bmatrix} = \frac{1}{\sqrt{2}}\begin{bmatrix} a_0\hat{e}_0 \\ \alpha_1\varepsilon_1 \end{bmatrix}, \tag{66}$$

$$\mathbf{P}_{10}|\boldsymbol{\Psi}_{ob}\rangle = \frac{1}{\sqrt{2}}\begin{bmatrix} a_1\begin{pmatrix}0\\1\end{pmatrix} \\ \alpha_0\begin{pmatrix}1\\0\end{pmatrix} \end{bmatrix} = \frac{1}{\sqrt{2}}\begin{bmatrix} a_1\hat{e}_1 \\ \alpha_0\varepsilon_0 \end{bmatrix}. \tag{67}$$

The multiplication of the crossed double projections (64) and the double projections (58) is given by

$$\boldsymbol{P}_{01}\boldsymbol{P}_0 = \boldsymbol{P}_0\boldsymbol{P}_{01} = \begin{bmatrix} \hat{P}_0 & \hat{0} \\ \hat{0} & \hat{0} \end{bmatrix} \equiv \mathbf{Q}_0, \qquad \boldsymbol{P}_{01}\boldsymbol{P}_1 = \boldsymbol{P}_1\boldsymbol{P}_{01} = \begin{bmatrix} \hat{0} & \hat{0} \\ \hat{0} & \hat{P}_1^{(\mu)} \end{bmatrix} \equiv \mathbf{Q}_1^{(\mu)}, \tag{68}$$

$$\boldsymbol{P}_{10}\boldsymbol{P}_0 = \boldsymbol{P}_0\boldsymbol{P}_{10} = \begin{bmatrix} \hat{0} & \hat{0} \\ \hat{0} & \hat{P}_0^{(\mu)} \end{bmatrix} \equiv \mathbf{Q}_0^{(\mu)}, \qquad \boldsymbol{P}_{10}\boldsymbol{P}_1 = \boldsymbol{P}_1\boldsymbol{P}_{10} = \begin{bmatrix} \hat{P}_1 & \hat{0} \\ \hat{0} & \hat{0} \end{bmatrix} \equiv \mathbf{Q}_1, \tag{69}$$

where the operators $\mathbf{Q}_0, \mathbf{Q}_1$ and $\mathbf{Q}_0^{(\mu)}, \mathbf{Q}_1^{(\mu)}$ satisfy

$$\mathbf{Q}_0^2 = \mathbf{Q}_0, \ \mathbf{Q}_1^2 = \mathbf{Q}_1, \ \mathbf{Q}_1\mathbf{Q}_0 = \mathbf{Q}_0\mathbf{Q}_1 = \mathbf{0}, \tag{70}$$

$$\mathbf{Q}_0^{(\mu)2} = \mathbf{Q}_0^{(\mu)}, \quad \mathbf{Q}_1^{(\mu)2} = \mathbf{Q}_1^{(\mu)}, \quad \mathbf{Q}_1^{(\mu)}\mathbf{Q}_0^{(\mu)} = \mathbf{Q}_0^{(\mu)}\mathbf{Q}_1^{(\mu)} = \mathbf{0}, \tag{71}$$

$$\mathbf{Q}_1^{(\mu)}\mathbf{Q}_0 = \mathbf{Q}_0^{(\mu)}\mathbf{Q}_1 = \mathbf{Q}_1\mathbf{Q}_0^{(\mu)} = \mathbf{Q}_0\mathbf{Q}_1^{(\mu)} = \mathbf{0}, \tag{72}$$

and we call these "half Kronecker (double) projections".

The relations above imply that the process of measurement when using Kronecker obscure qubits (i.e. for quantum computation with truth or membership) is more complicated than in the standard case.

To show this, let us calculate the "obscure" analogs of expected values for the projections above. Using the notation

$$\overline{\mathbf{A}} \equiv \langle \mathbf{\Psi}_{ob} | \mathbf{A} | \mathbf{\Psi}_{ob} \rangle. \tag{73}$$

Then, using (43)–(45) for the projection operators P_i, P_{ij}, Q_i, $Q_i^{(\mu)}$, $i, j = 0, 1, i \neq j$, we obtain (cf. (39))

$$\overline{P}_i = \frac{|a_i|^2 + \alpha_i^2}{2}, \qquad \overline{P}_{ij} = \frac{|a_i|^2 + \alpha_j^2}{2}, \tag{74}$$

$$\overline{Q}_i = \frac{|a_i|^2}{2}, \qquad \overline{Q}_i^{(\mu)} = \frac{\alpha_i^2}{2}. \tag{75}$$

So follows the relation between the "obscure" analogs of expected values of the projections

$$\overline{P}_i = \overline{Q}_i + \overline{Q}_i^{(\mu)}, \qquad \overline{P}_{ij} = \overline{Q}_i + \overline{Q}_j^{(\mu)}. \tag{76}$$

Taking the "ket" corresponding to the "bra" Kronecker qubit (52) in the form

$$\langle \mathbf{\Psi}_{ob} | - \frac{1}{\sqrt{2}} [a_0^* (1 \quad 0), \quad a_0 (1 \quad 0)] + \frac{1}{\sqrt{2}} [a_1^* (0 \quad 1), \quad a_1 (0 \quad 1)], \tag{77}$$

a Kronecker (4×4) obscure analog of the density matrix for a pure state is given by (cf. (42))

$$\rho_{ob}^{(2)} = |\mathbf{\Psi}_{ob}\rangle \langle \mathbf{\Psi}_{ob}| = \frac{1}{2} \begin{pmatrix} |a_0|^2 & a_0 a_1^* & a_0 \alpha_0 & a_0 \alpha_1 \\ a_1 a_0^* & |a_1|^2 & a_1 \alpha_0 & a_1 \alpha_1 \\ \alpha_0 a_0^* & \alpha_0 a_1^* & \alpha_0^2 & \alpha_0 \alpha_1 \\ \alpha_1 a_0^* & \alpha_1 a_1^* & \alpha_0 \alpha_1 & \alpha_1^2 \end{pmatrix}. \tag{78}$$

If the Born rule for the membership functions (13) and the conditions (14)–(15) are satisfied, the density matrix (78) is non-invertible, because $\det \rho_{ob}^{(2)} = 0$ and has unit trace $\mathrm{tr}\, \rho_{ob}^{(2)} = 1$, but is not idempotent $\left(\rho_{ob}^{(2)}\right)^2 \neq \rho_{ob}^{(2)}$ (as it holds for the ordinary quantum density matrix [1]).

7. Kronecker obscure-quantum gates

In general, (double) "obscure-quantum computation" with L Kronecker obscure qubits (or qudits) can be performed by a product of unitary (block) matrices $\boldsymbol{U}$ of the (double size to the standard one) size $2 \times \left(2^L \times 2^L\right)$ (or $2 \times \left(n^L \times n^L\right)$), $\boldsymbol{U}^\dagger \boldsymbol{U} = \boldsymbol{I}$ (here $\boldsymbol{I}$ is the unit matrix of the same size as $\boldsymbol{U}$). We can also call such computation a "quantum computation with truth" (or with membership).

Let us consider obscure-quantum computation with one Kronecker obscure qubit. Informally, we can present the Kronecker obscure qubit (52) in the form

$$|\mathbf{\Psi}_{ob}\rangle = \begin{bmatrix} \frac{1}{\sqrt{2}}\begin{pmatrix} a_0 \\ a_1 \end{pmatrix} \\ \frac{1}{\sqrt{2}}\begin{pmatrix} \alpha_0 \\ \alpha_1 \end{pmatrix}^{(\mu)} \end{bmatrix}. \tag{79}$$

Thus, the state $|\mathbf{\Psi}_{ob}\rangle$ can be interpreted as a "vector" in the direct product (not tensor product) space $\mathcal{H}_q^{(2)} \times \mathcal{V}_{memb}^{(2)}$, where $\mathcal{H}_q^{(2)}$ is the standard two-dimenional Hilbert space of the qubit, and $\mathcal{V}_{memb}^{(2)}$ can be treated as the "membership space" which has a different nature from the qubit space and can have a more complex structure. For discussion of such spaces, see, e.g. [5, 6, 8, 9]. In general, one can consider obscure-quantum computation as a set of abstract computational rules, independently of the introduction of the corresponding spaces.

An obscure-quantum gate will be defined as an elementary transformation on an obscure qubit (79) and is performed by unitary (block) matrices of size 4×4 (over $\mathbb{C}$) acting in the total space $\mathcal{H}_q^{(2)} \times \mathcal{V}_{memb}^{(2)}$

$$\boldsymbol{U} = \begin{pmatrix} \hat{U} & \hat{0} \\ \hat{0} & \hat{U}^{(\mu)} \end{pmatrix}, \qquad \boldsymbol{U}\boldsymbol{U}^{\dagger} = \boldsymbol{U}^{\dagger}\boldsymbol{U} = \boldsymbol{I}, \tag{80}$$

$$\hat{U}U^{\dagger} = \hat{U}^{\dagger}\hat{U} = \hat{I}, \quad \hat{U}^{(\mu)}\hat{U}^{(\mu)\dagger} = \hat{U}^{(\mu)\dagger}\hat{U}^{(\mu)} = \hat{I}, \qquad \hat{U} \in \operatorname{End}\mathcal{H}_q^{(2)}, \hat{U}^{(\mu)} \in \operatorname{End}\mathcal{V}_{memb}^{(2)}, \tag{81}$$

where $\boldsymbol{I}$ is the unit 4×4 matrix, $\hat{I}$ is the unit 2×2 matrix, $\hat{U}$ and $\hat{U}^{(\mu)}$ are unitary 2×2 matrices acting on the probability and membership "subspaces" respectively. The matrix $\hat{U}$ (over $\mathbb{C}$) will be called a quantum gate, and we call the matrix $\hat{U}^{(\mu)}$ (over $\mathbb{R}$) an "obscure gate". We assume that the obscure gates $\hat{U}^{(\mu)}$ are of the same shape as the standard quantum gates, but they act in the other (membership) space and have only real elements (see, e.g. [1]). In this case, an obscure-quantum gate is characterized by the pair $\left\{\hat{U}, \hat{U}^{(\mu)}\right\}$, where the components are known gates (in various combinations), e.g., for one qubit gates: `Hadamard`, `Pauli-X` (NOT),Y,Z (or two qubit gates e.g. CNOT, SWAP, etc.). The transformed qubit then becomes (informally)

$$\mathbf{U}|\mathbf{\Psi}_{ob}\rangle = \begin{bmatrix} \frac{1}{\sqrt{2}}\hat{U}\begin{pmatrix} a_0 \\ a_1 \end{pmatrix} \\ \frac{1}{\sqrt{2}}\hat{U}^{(\mu)}\begin{pmatrix} \alpha_0 \\ \alpha_1 \end{pmatrix}^{(\mu)} \end{bmatrix}. \tag{82}$$

Thus the quantum and the membership parts are transformed independently for the block diagonal form (80). Some examples of this can be found, e.g., in [36, 37, 45]. Differences between the parts were mentioned in [46]. In this case, an obscure-quantum network is "physically" realised by a device performing elementary operations in sequence on obscure qubits (by a product of matrices), such that the quantum and membership parts are synchronized in time (for a discussion of the obscure part of such physical devices, see [19, 20, 47, 48]). Then, the result of the obscure-quantum computation consists of the quantum probabilities of the states together with the calculated "level of truth" for each of them (see, e.g. [18]).

For example, the obscure-quantum gate $\boldsymbol{U}_{\hat{H},\mathrm{NOT}} = \{\mathtt{Hadamard}, \mathrm{NOT}\}$ acts on the state $\hat{E}_0$ (55) as follows

$$\boldsymbol{U}_{\hat{H},\mathrm{NOT}}\hat{E}_0 = \boldsymbol{U}_{\hat{H},\mathrm{NOT}} \begin{bmatrix} \begin{pmatrix} 1 \\ 0 \end{pmatrix} \\ \begin{pmatrix} 1 \\ 0 \end{pmatrix}^{(\mu)} \end{bmatrix} = \begin{bmatrix} \frac{1}{\sqrt{2}} \begin{pmatrix} 1 \\ 1 \end{pmatrix} \\ \begin{pmatrix} 0 \\ 1 \end{pmatrix}^{(\mu)} \end{bmatrix} = \begin{bmatrix} \frac{1}{\sqrt{2}}(\hat{e}_0 + \hat{e}_1) \\ \varepsilon_1 \end{bmatrix}. \tag{83}$$

It would be interesting to consider the case when $\boldsymbol{U}$ (80) is not block diagonal and try to find possible "physical" interpretations of the non-diagonal blocks.

8. Double entanglement

Let us introduce a register consisting of two obscure qubits ($L = 2$) in the computational basis $|\hat{i}\hat{j}'\rangle = |\hat{i}\rangle \otimes |\hat{j}'\rangle$ as follows

$$\left|\boldsymbol{\Psi}_{ob}^{(n=2)}(L=2)\right\rangle = |\boldsymbol{\Psi}_{ob}(2)\rangle = \frac{\hat{B}_{00'}|\hat{0}\hat{0}'\rangle + \hat{B}_{10'}|\hat{1}\hat{0}'\rangle + \hat{B}_{01'}|\hat{0}\hat{1}'\rangle + \hat{B}_{11'}|\hat{1}\hat{1}'\rangle}{\sqrt{2}}, \tag{84}$$

determined by two-dimensional "vectors" (encoding obscure-quantum amplitudes)

$$\hat{B}_{ij'} = \begin{bmatrix} b_{ij'} \\ \beta_{ij'} \end{bmatrix}, \qquad i,j = 0,1, \qquad j' = 0',1', \tag{85}$$

where $b_{ij'} \in \mathbb{C}$ are probability amplitudes for a set of pure states and $\beta_{ij'} \in \mathbb{R}$ are the corresponding membership amplitudes. By analogy with (43) and (46) the normalization factor in (84) is chosen so that

$$\langle \boldsymbol{\Psi}_{ob}(2) | \boldsymbol{\Psi}_{ob}(2) \rangle = 1, \tag{86}$$

if (cf. (14)–(15))

$$|b_{00'}|^2 + |b_{10'}|^2 + |b_{01'}|^2 + |b_{11'}|^2 = 1, \tag{87}$$

$$\beta_{00'}^2 + \beta_{10'}^2 + \beta_{01'}^2 + \beta_{11'}^2 = 1. \tag{88}$$

A state of two qubits is "entangled", if it cannot be decomposed as a product of two one-qubit states, and otherwise it is "separable" (see, e.g. [1]). We define a product of two obscure qubits (43) as

$$|\boldsymbol{\Psi}_{ob}\rangle \otimes |\boldsymbol{\Psi}'_{ob}\rangle = \frac{\hat{A}_0 \otimes_H \hat{A}'_0|\hat{0}\hat{0}'\rangle + \hat{A}_1 \otimes_H \hat{A}'_0|\hat{1}\hat{0}'\rangle + \hat{A}_0 \otimes_H \hat{A}'_1|\hat{0}\hat{1}'\rangle + \hat{A}_1 \otimes_H \hat{A}'_1|\hat{1}\hat{1}'\rangle}{2}, \tag{89}$$

where $\otimes_H$ is the Hadamard product (53). Comparing (84) and (89) we obtain two sets of relations, for probability amplitudes and for membership amplitudes

$$b_{ij'} = \frac{1}{\sqrt{2}} a_i a_{j'}, \tag{90}$$

$$\beta_{ij'} = \frac{1}{\sqrt{2}} \alpha_i \alpha_{j'}, \qquad i,j = 0,1, \qquad j' = 0', 1'. \tag{91}$$

In this case, the relations (14)–(15) give (87)–(88).

Two obscure-quantum qubits are entangled, if their joint state (84) cannot be presented as a product of one qubit states (89), and in the opposite case the states are called totally separable. It follows from (90)–(91), that there are two general conditions for obscure qubits to be entangled

$$b_{00'} b_{11'} \neq b_{10'} b_{01'}, \qquad \text{or } \det \boldsymbol{b} \neq 0, \; \boldsymbol{b} = \begin{pmatrix} b_{00'} & b_{01'} \\ b_{10'} & b_{11'} \end{pmatrix}, \tag{92}$$

$$\beta_{00'} \beta_{11'} \neq \beta_{10'} \beta_{01'}, \qquad \text{or } \det \boldsymbol{\beta} \neq 0, \; \boldsymbol{\beta} = \begin{pmatrix} \beta_{00'} & \beta_{01'} \\ \beta_{10'} & \beta_{11'} \end{pmatrix}. \tag{93}$$

The first Eq. (92) is the entanglement relation for the standard qubit, while the second condition (93) is for the membership amplitudes of the two obscure qubit joint state (84). The presence of two different conditions (92)–(93) leads to new additional possibilities (which do not exist for ordinary qubits) for "partial" entanglement (or "partial" separability), when only one of them is fulfilled. In this case, the states can be entangled in one subspace (quantum or membership) but not in the other.

The measure of entanglement is numerically characterized by the concurrence. Taking into account the two conditions (92)–(93), we propose to generalize the notion of concurrence for two obscure qubits in two ways. First, we introduce the "vector obscure concurrence"

$$\hat{C}_{vect} = \begin{bmatrix} C_q \\ C^{(\mu)} \end{bmatrix} = 2 \begin{bmatrix} |\det \boldsymbol{b}| \\ |\det \beta| \end{bmatrix}, \tag{94}$$

where $\boldsymbol{b}$ and β are defined in (92)–(93), and $0 \le C_q \le 1$, $0 \le C^{(\mu)} \le 1$. The corresponding "scalar obscure concurrence" can be defined as

$$C_{scal} = \sqrt{\frac{|\det \boldsymbol{b}|^2 + |\det \beta|^2}{2}}, \tag{95}$$

such that $0 \le C_{scal} \le 1$. Thus, two obscure qubits are totally separable, if $C_{scal} = 0$. For instance, for an obscure analog of the (maximally entangled) Bell state

$$|\boldsymbol{\Psi}_{ob}(2)\rangle = \frac{1}{\sqrt{2}} \left(\begin{bmatrix} \frac{1}{\sqrt{2}} \\ \frac{1}{\sqrt{2}} \end{bmatrix} |\hat{0}0'\rangle + \begin{bmatrix} \frac{1}{\sqrt{2}} \\ \frac{1}{\sqrt{2}} \end{bmatrix} |\hat{1}1'\rangle \right) \tag{96}$$

we obtain

$$\hat{C}_{vect} = \begin{bmatrix} 1 \\ 1 \end{bmatrix}, \qquad C_{scal} = 1. \tag{97}$$

A more interesting example is the "intermediately entangled" two obscure qubit state, e.g.

$$|\boldsymbol{\Psi}_{ob}(2)\rangle = \frac{1}{\sqrt{2}}\left(\begin{bmatrix}\frac{1}{2}\\ \frac{1}{\sqrt{2}}\end{bmatrix}|\hat{0}0'\rangle + \begin{bmatrix}\frac{1}{4}\\ \frac{\sqrt{5}}{4}\end{bmatrix}|\hat{1}0'\rangle + \begin{bmatrix}\frac{\sqrt{3}}{4}\\ \frac{1}{2\sqrt{2}}\end{bmatrix}|\hat{0}1'\rangle + \begin{bmatrix}\frac{1}{\sqrt{2}}\\ \frac{1}{4}\end{bmatrix}|\hat{1}1'\rangle\right), \quad (98)$$

where the amplitudes satisfy (87)–(88). If the Born-like rule (as in (13)) holds for the membership amplitudes, then the probabilities and membership functions of the states in (98) are

$$p_{00'} = \frac{1}{4}, \quad p_{10'} = \frac{1}{16}, \quad p_{01} = \frac{3}{16}, \quad p_{11'} = \frac{1}{2}, \quad (99)$$

$$\mu_{00'} = \frac{1}{2}, \quad \mu_{10'} = \frac{5}{16}, \quad \mu_{01'} = \frac{1}{8}, \quad \mu_{11'} = \frac{1}{16}. \quad (100)$$

This means that, e.g., the state $|\hat{1}0'\rangle$ will be measured with the quantum probability 1/16 and the membership function ("truth" value) 5/16. For the entangled obscure qubit (98) we obtain the concurrences

$$\hat{C}_{vect} = \begin{bmatrix}\frac{1}{2}\sqrt{2} - \frac{1}{8}\sqrt{3}\\ \frac{1}{8}\sqrt{2}\sqrt{5} - \frac{1}{4}\sqrt{2}\end{bmatrix} = \begin{bmatrix}0.491\\ 0.042\end{bmatrix}, \quad C_{scal} = \sqrt{\frac{53}{128} - \frac{1}{16}\sqrt{5} - \frac{1}{16}\sqrt{2}\sqrt{3}}$$
$$= 0.348. \quad (101)$$

In the vector representation (49)–(52) we have

$$|\hat{i}j'\rangle = |\hat{i}\rangle \otimes |\hat{j}'\rangle = \begin{bmatrix}\hat{e}_i \otimes_K \hat{e}_{j'}\\ \varepsilon_i \otimes_K \varepsilon_{j'}\end{bmatrix}, \quad i,j = 0,1, \quad j' = 0',1', \quad (102)$$

where $\otimes_K$ is the Kronecker product (48), and $\hat{e}_i$, ε_i are defined in (50)–(51). Using (85) and the Kronecker-like product (49), we put (informally, with no summation)

$$\hat{B}_{ij'}|\hat{i}j'\rangle = \begin{bmatrix}b_{ij'}\hat{e}_i \otimes_K \hat{e}_{j'}\\ \beta_{ij'}\varepsilon_i \otimes_K \varepsilon_{j'}\end{bmatrix}, \quad i,j = 0,1, \quad j' = 0',1'. \quad (103)$$

To clarify our model, we show here a manifest form of the two obscure qubit state (98) in the vector representation

$$|\boldsymbol{\Psi}_{ob}(2)\rangle = \frac{1}{\sqrt{2}}\left(\begin{bmatrix}\frac{1}{2}\begin{pmatrix}1\\0\\1\\0\end{pmatrix}\\ \frac{1}{\sqrt{2}}\begin{pmatrix}1\\0\\1\\0\end{pmatrix}^{(\mu)}\end{bmatrix} + \begin{bmatrix}\frac{1}{4}\begin{pmatrix}0\\1\\1\\0\end{pmatrix}\\ \frac{\sqrt{5}}{4}\begin{pmatrix}0\\1\\1\\0\end{pmatrix}^{(\mu)}\end{bmatrix} + \begin{bmatrix}\frac{\sqrt{3}}{4}\begin{pmatrix}1\\0\\0\\1\end{pmatrix}\\ \frac{1}{2\sqrt{2}}\begin{pmatrix}1\\0\\0\\1\end{pmatrix}^{(\mu)}\end{bmatrix} + \begin{bmatrix}\frac{1}{\sqrt{2}}\begin{pmatrix}0\\1\\0\\1\end{pmatrix}\\ \frac{1}{4}\begin{pmatrix}0\\1\\0\\1\end{pmatrix}^{(\mu)}\end{bmatrix}\right). \quad (104)$$

The states above may be called "symmetric two obscure qubit states". However, there are more general possibilities, as may be seen from the r.h.s. of (103) and (104), when the indices of the first and second rows do not coincide. This would allow more possible states, which we call "non-symmetric two obscure qubit states". It would be worthwhile to establish their possible physical interpretation.

The above constructions show that quantum computing using Kronecker obscure qubits can involve a rich structure of states, giving a more detailed description with additional variables reflecting vagueness.

9. Conclusions

We have proposed a new scheme for describing quantum computation bringing vagueness into consideration, in which each state is characterized by a "measure of truth" Ȧ membership amplitude is introduced in addition to the probability amplitude in order to achieve this, and we are led thereby to the concept of an obscure qubit. Two kinds of these are considered: the "product" obscure qubit, in which the total amplitude is the product of the quantum and membership amplitudes, and the "Kronecker" obscure qubit, where the amplitudes are manipulated separately. In latter case, the quantum part of the computation is based, as usual, in Hilbert space, while the "truth" part requires a vague/fuzzy set formalism, and this can be performed in the framework of a corresponding fuzzy space. Obscure-quantum computation may be considered as a set of rules (defining obscure-quantum gates) for managing quantum and membership amplitudes independently in different spaces. In this framework we obtain not only the probabilities of final states, but also their membership functions, i.e. how much "trust" we should assign to these probabilities. Our approach considerably extends the theory of quantum computing by adding the logic part directly to the computation process. Future challenges could lie in the direction of development of the corresponding logic hardware in parallel with the quantum devices.

Acknowledgements

The first author (S.D.) is deeply thankful to Geoffrey Hare and Mike Hewitt for thorough language checking.

Author details

Steven Duplij* and Raimund Vogl
Center for Information Technology (WWU IT), University of Münster, Münster, Deutschland

*Address all correspondence to: douplii@uni-muenster.de

References

[1] Nielsen, M.A.; Chuang, I.L. *Quantum Computation and Quantum Information*; Cambridge University Press: Cambridge, 2000; p. 676.

[2] Kaye, P.; Laflamme, R.; Mosca, M. *An introduction to quantum computing*; Oxford Univ. Press: Oxford, 2007.

[3] Williams, C.P.; Clearwater, S.H. *Explorations in Quantum Computing*; Springer-Verlag: New York-Berlin-Heidelberg, 1998.

[4] Goodman, I.R.; Nguyen, H.T. Fuzziness and randomness. Statistical Modeling, Analysis and Management of Fuzzy Data; Bertoluzza, C.; Gil, M.A.; Ralescu, D.A., Eds. Springer, 2002, pp. 3–21. doi: 10.1007/978-3-7908-1800-0_1.

[5] Dubois, D.; Nguyen, H.T.; Prade, H. Possibility Theory, Probability and Fuzzy Sets, Misunderstandings, Bridges and Gaps. Fundamentals of Fuzzy Sets; Dubois, D.; Prade, H., Eds.; Springer: Boston, 2000; pp. 343–438. doi: 10.1007/978-1-4615-4429-6_8.

[6] Bělohlávek, R. *Fuzzy relational systems: Foundations and principles*; Springer: New York, 2002.

[7] Dubois, D.; Prade, H., Eds. *Fundamentals of fuzzy sets*; Vol. 7, *The Handbooks of Fuzzy Sets Series*, Kluwer: Boston, 2000; p. 647. doi: 10.1007/978-1-4615-4429-6.

[8] Smith, N.J.J. *Vagueness and degrees of truth*; Oxford University Press: Oxford, 2008.

[9] Zimmermann, H.J. *Fuzzy set theory and its applications*, second ed.; Springer: New York, 2011.

[10] Pykacz, J. *Quantum Physics, Fuzzy Sets and Logic*; Springer: Cham, 2015. doi: 10.1007/978-3-319-19384-7.

[11] Dvurečenskij, A.; Chovanec, F. Fuzzy quantum spaces and compatibility. *Int. J. Theor. Phys.* **1988**, *27*, 1069–1082. doi: 10.1007/bf00674352.

[12] Bartková, R.; Riečan, B.; Tirpáková, A. *Probability Theory for Fuzzy Quantum Spaces with Statistical Applications*; Bentham Science Publishers: Sharjah, 2017. doi: 10.2174/97816810853881170100006.

[13] Granik, A. Fuzziness in quantum mechanics. Adaptive Computing: Mathematics, Electronics, and Optics: A Critical Review; Chen, S.S.; Caulfield, H.J., Eds. International Society for Optics and Photonics, SPIE, 1994, Vol. 10277, pp. 228 – 249. doi: 10.1117/12.171191.

[14] Gudder, S. A theory of amplitudes. *J. Math. Phys.* **1988**, *29*, 2020–2035. doi: 10.1063/1.527860.

[15] Gudder, S. Fuzzy quantum probability theory. Foundations of probability and physics — 3; Khrennikov, A., Ed.; Amer. Inst. Phys.: Melville, 2005; Vol. 750, *AIP Conf. Proc.*, pp. 114–121. doi: 10.1063/1.1874565.

[16] Gudder, S. Fuzzy amplitude densities and stochastic quantum mechanics. *Foundations of Physics* **1989**, *19*, 293–317. doi: 10.1007/BF00734561.

[17] Gudder, S. Partial Hilbert spaces and amplitude functions. *Ann. Ins. Henri Poincaré* **1986**, *45*, 311–326.

[18] Bolotin, A. Truth values of quantum phenomena. *Int. J. Theor. Phys.* **2018**, *57*, 2124–2132.

[19] Hirota, K.; Ozawa, K. The concept of fuzzy flip-flop. *IEEE Transactions on Systems, Man, and Cybernetics* **1989**, *19*, 980–997. doi: 10.1109/21.44013.

[20] Virant, J. *Design considerations of time in fuzzy systems*; Vol. 35, Kluwer: Dordrecht, 2000; pp. xxxvi+475. doi: 10.1007/978-1-4615-4673-3.

[21] Buniy, R.V.; Hsu, S.D.; Zee, A. Is Hilbert space discrete? *Phys. Lett.* **2005**, *B630*, 68–72. doi: https://doi.org/10.1016/j.physletb.2005.09.084.

[22] Marchiolli, M.A.; Ruzzi, M. Theoretical formulation of finite-dimensional discrete phase spaces: I. Algebraic structures and uncertainty principles. *Annals of Physics* **2012**, *327*, 1538–1561. doi: 10.1016/j.aop.2012.02.015.

[23] Barrett, J. Information processing in generalized probabilistic theories. *Phys. Rev. A*, *75*, 032304.

[24] Müller, M. Does Probability become Fuzzy in Small Regions of Spacetime? *Phys. Lett.* **2009**, *B673*, 166–167. doi: 10.1016/j.physletb.2009.02.017.

[25] Brylinski, J.L.; Brylinski, R. Universal quantum gates. Mathematics of Quantum Computation; Chapman & Hall/CRC Press: Boca Raton, 1994; pp. 124–134.

[26] Rungta, P.; Munro, W.J.; Nemoto, K.; Deuar, P.; Milburn, G.J.; Caves, C.M. Qudit Entanglement. Directions in Quantum Optics; Carmichael, H.J.; Glauber, R.J.; Scully, M.O., Eds.; Springer-Verlag: Berlin-Heidelberg, 2001; Vol. 561, pp. 149–164. doi: 10.1007/3-540-40894-0_14.

[27] Kimura, G. The Bloch Vector for N-Level Systems. *Phys. Lett.*, *A314*, 339–349.

[28] Genovese, M.; Traina, P. Review on qudits production and their application to Quantum Communication and Studies on Local Realism. *Adv. Sci. Lett.* **2008**, *1*, 153–160.

[29] Wang, Y.; Hu, Z.; Sanders, B.C.; Kais, S. Qudits and high-dimensional quantum computing. arxiv:2008.00959.

[30] Deutsch, D. Quantum theory, the Church-Turing principle and the universal quantum computer. *Proc. Roy. Soc. London* **1985**, *A400*, 96–117.

[31] Barenco, A.; Bennett, C.H.; Cleve, R.; DiVincenzo, D.P.; Margolus, N.; Shor, P.; Sleator, T.; Smolin, J.A.; Weinfurter, H. Elementary gates for quantum computation. *Phys. Rev.* **1995**, *52*, 3457–3467.

[32] Lloyd, S. Almost Any Quantum Logic Gate is Universal. *Phys. Rev. Lett.* **1995**, *75*, 346–349.

[33] Buckley, J.J. Fuzzy complex numbers. *Fuzzy Sets Syst.* **1989**, *33*, 333–345.

[34] Ramot, D.; Milo, R.; Friedman, M.; Kandel, A. Complex fuzzy sets. *IEEE Transactions on Fuzzy Systems* **2002**, *10*, 171–186. doi: 10.1109/91.995119.

[35] Garrido, A. Axiomatic of Fuzzy Complex Numbers. *Axioms* **2012**, *1*, 21–32. doi: 10.3390/axioms1010021.

[36] Mannucci, M.A. Quantum Fuzzy Sets: Blending Fuzzy Set Theory and Quantum Computation. arXiv:cs/0604064.

[37] Maron, A.; Visintin, L.; Reiser, R.H.S.; Abeijon, A.M. Fuzzy computing from quantum computing–case study in Reichenbach implication class. *Mathware and Soft Comp. Mag.* **2013**, *20*, 86–93.

[38] Hannachi, M.S.; Hatakeyama, Y.; Hirota, K. Emulating Qubits with Fuzzy Logic. *J. Adv. Comput. Intel. Inf.* **2007**, *11*, 242–249. doi: 10.20965/jaciii.2007.p0242.

[39] Rybalov, A.; Kagan, E.; Rapoport, A.; Ben-Gal, I. Fuzzy Implementation of Qubits Operators. *J. Comput. Sci. Syst. Biol.* **2014**, *07*, 163–168. doi: 10.4172/jcsb.1000151.

[40] Battilotti, G.; Zizzi, P. Logical interpretation of a reversible measurement in quantum computing. quant-ph/0408068.

[41] Zizzi, P. Qubits and quantum spaces. *Int. J. Quant. Inf.* **2005**, *3*, 287–291.

[42] Zizzi, P.; Pessa, E. From $SU(2)$ gauge theory to qubits on the fuzzy sphere. *Int. J. Theor. Phys.* **2014**, *53*, 257–270.

[43] Cunha, R.A.F.; Sharma, N.; Toffano, Z.; Dubois, F. Fuzzy Logic Behavior of Quantum-Controlled Braitenberg Vehicle Agents. International Symposium on Quantum Interaction (QI 2018); Coecke, B.; Lambert-Mogiliansky, A., Eds.; Springer: Berlin, 2019; pp. 111–122.

[44] Toffano, Z.; Dubois, F. Eigenlogic: Interpretable Quantum Observables with applications to Fuzzy Behavior of Vehicular Robots. Science with and for Society: Contributions of Cybernetics and Systems; World Organisation of Systems and Cybernetics, , 2017; pp. 1–5.

[45] Domenech, G.; Freytes, H. Fuzzy propositional logic associated with quantum computational gates. *Int. J. Theor. Phys.* **2006**, *45*, 237–270. doi: 10.1007/s10773-005-9019-6.

[46] Kreinovich, V.; Kohout, L.; Kim, E. Square root of 'not': a major difference between fuzzy and quantum logics. *Int. J. General Systems* **2011**, *40*, 111–127.

[47] Kóczy, L.T.; Hirota, K. Digital circuits based on algebraic fuzzy operations. Progress in fuzzy sets and systems (Vienna, 1989); Janko, W.H.; Roubens, M.; Zimmermann, H.J., Eds.; Kluwer: Dordrecht, 1990; Vol. 5, pp. 100–114.

[48] Kosko, B. *Fuzzy engineering*; Prentice Hall, Inc.: Upper Saddle River, 1997; pp. xxvi+549.

Chapter 7

Device Independence and the Quest towards Physical Limits of Privacy

Gopalan Raghavan

Abstract

There is a looming threat over current methods of data encryption through advances in quantum computation. Interestingly, this potential threat can be countered through the use of quantum resources such as coherent superposition, entanglement and inherent randomness. These, together with non-clonability of arbitrary quantum states, offer provably secure means of sharing encryption keys between two parties. This physically assured privacy is however provably secure only in theory but not in practice. Device independent approaches seek to provide physically assured privacy of devices of untrusted origin. The quest towards realization of such devices is predicated on conducting loop-hole-free Bell tests which require the use of certified quantum random number generators. The experimental apparatuses for conducting such tests themselves use non-ideal sources, detectors and optical components making such certification extremely difficult. This expository chapter presents a brief overview (not a review) of Device Independence and the conceptual and practical difficulties it entails.

Keywords: QRNG, QKD, device independence, loop-hole-free Bell tests, nonlocality

1. Introduction

The advent of quantum technologies holds the promise of novel innovations in computing, communication and sensing. Quintessential quantum properties such as superposition and entanglement are perceived as essential resources for the realization of these technologies. Quantum states allow for non-local correlations under no-signaling conditions [1, 2]. Claims to "quantum supremacy" in quantum computing or provable security in quantum cryptography hinges on the assertion that quantum resources are not only needed, but can be gainfully employed for realizing functionalities, which classical resources cannot supply. Quantum cryptography or rather quantum key distribution (QKD) schemes are claimed to be provably secure based on the quantum nature of the carriers of information. The claim on information security relies on the fact, that perfect copies of arbitrary quantum states cannot be made (the "no-cloning" theorem) [3] and the fact that measurements disturb the state of the system in an irreversible fashion, resulting in perfectly random outcomes. In quantum key distribution protocols such as the Ekert's [4], non-local correlations between a pair of entangled photons are utilized to realize secure key exchange between two parties in space-like separated regions. One of the key components of a QKD device is a quantum random number generator (QRNG) [5–7]. These devices are believed to generate perfect and inherently random sequences that cannot be produced by any device based on classical phenomena or by using mathematical algorithms however complex they might be. High-speed QRNGs are an important requirement not only for QKD but have potential uses in gambling, commerce, and classical cryptography. Given the importance of this device for cryptography, one should test this device before using it. When a consumer buys a piece of quantum-enabled equipment such as QKD boxes or QRNGs, there is a need to find out whether it is the "real McCoy" and the hardware performs as advertised. For instances, the QRNG, sourced through an untrusted party may generate a seemingly random sequences on demand, but it begs the question, whether these sequences arise from a genuine quantum process and have not been generated through some classical or algorithmic means? Alternately, the supplier of the device could have generated a very large sequence through a quantum process and stored it in the device while retaining copy for herself. Even without assuming any evil intent on part of the supplier, the device could also be unreliable because of imperfections in the source or detectors or even due the noise being well-above permissible thresholds. Standard statistical tests for randomness such as DIEHARD and DIEHARDER provided by NIST, USA [8, 9] are not the solution to this problem. Statistical tests for randomness merely certify the absence of certain patterns in the sequences within the bounds of finite computational power at the disposal of the of the user. It would be logical fallacy to think that absence of evidence is evidence of absence. Statistical tests are therefore not tests of genuine quantum randomness and most certainly do not provide any assurance regarding the privacy of the data that is generated. For the QRNG to satisfy its claimed performance not only should the output of the device be perfectly random to the user, but to any observer including the supplier of device. Then, and only then, could a "QRNG" be deemed to be a Perfect and Private Random Number generator or PPRNG as we shall call it henceforth. The advantages of quantum resources for secure communications or random number generation are a theoretical fact but, demonstrating that a piece of hardware actually exploits quantum properties of matter and fields in an effective manner is a non-trivial problem. The issue at hand is an important one because, it is related to very reliability of the quantum device itself. Extraordinary, though it seems, it is possible that a certain class of QRNGs and QKDs of illicit and unknown provenance, could be certified to be provably secure through the performance of certain class of tests called Bell tests performed on them [10–13]. Such Bell test certified devices are however extremely difficult to realize and currently the rates of random number generation with them are extremely small. Before we get into issues of device independence, we first examine some aspects of randomness, non-locality and non-local correlations.

2. Randomness, nonlocality and non-local correlations

The Famous paper by Einstein, Podolsky and Rosen in Physical Review (1935) [14] was a watershed event setting-off a vigorous debate on the so-called hidden variable theories. Their central conclusion was that quantum mechanical description of physical systems is incomplete and that, quantum mechanical rules must be supplemented with additional variables to exorcize the seemingly inherent randomness of nature. Bohr rebuts these arguments in Physical Review [15] claiming that quantum mechanics deals with the statistical outcomes of the interactions of a microscopic system with a macroscopic classical apparatus and nothing further. In physical theories, classical or quantum, every system is associated with a mathematical description of it called the state. Given the state of a system at a certain instant of time, the time evolution of the system is described by Newton's laws of motion in classical mechanics and the Schrödinger equation in quantum mechanics respectively. Both these equations are perfectly deterministic and reversible in time. The time evolution of quantum states is described is unitary. In classical systems, randomness arises primarily due to inadequate information about the initial conditions especially when the degrees of freedom are extremely large. Quite often, we may also choose to ignore vast amounts of data simply because we do not have the computational resources to handle it. This sort of coarse graining of data becomes a practical expediency. The solution to handling full-blown turbulence by solving the Navier–Stokes equation would come to mind. In systems exhibiting classical chaos [16], even though the underlying equation of motion are deterministic, the apparent randomness and unpredictability arises because of sensitivity to initial conditions and the finite precision with which the initial conditions are supplied. In summary, classical randomness arises from ignorance or computational limitations and is therefore not an inherent property of nature. Such randomness is deemed to be epistemological in character. In quantum mechanics, while evolution itself is unitary, the outcomes of measurements performed on the quantum state are probabilistic. The expectation value of physical properties associated with observable $\hat{A}$ for an ensemble of measurements is given by the Born rule $\langle \hat{A} \rangle = \text{Tr}\left(\hat{\rho}\hat{A}\right)$. Given an ensemble of measurements M performed on identically prepared states, the probability of the jth outcome over a set of possible outcomes $\{E_i\}$ is given by $\left(p_j|M\right) = Tr(\hat{\rho}E_j)$ [2]. The Born rule thus provides us the leeway that links the abstract quantum state with observed phenomena. The outcome of single observation measurement is believed to be completely and inherently unpredictable and is an essential aspect of nature herself. Quantum mechanics does not offer any clue as to the physical origins of the observed randomness of outcomes. The implicit assumption is that the God of all things does play dice and is indeed an inveterate and compulsive gambler. Quantum randomness is therefore said to be inherent or ontological (ontic) randomness. It would be wise to bear in mind that there is no finality to any of these assertions. They are provisional to way we understand nature as of now. It is in this context; we should make a distinction between ontic and epistemic randomness; Ontic randomness is intrinsically associated with observable quantities of the system and related to self-adjoint operators. One or the other of the possible eigenvalues of these operators are manifest upon an observation. No à priori value can be associated with properties of the system. It is in this sense that one asserts that "quantum phenomena" are not realistic. Robust average values can however be assigned to average or expectation values of observables, which is what the Born formula helps us compute. Quantum mechanics is then an ensemble theory which provides us with recipe to calculate averages of repeated measurements made on the system. There is rich literature regarding the nature quantum state, the wavefunction itself. There are interpretations of quantum mechanics that significantly differ in their viewpoint. Since our purpose here is not to get deeply mired into foundations of quantum mechanics, we shall desist from such digressions, given the limited ambitions of this chapter.

It is a something of a fundamental theorem that purely local operations performed on single device, cannot be used to establish that the random sequences emitted by a given device has not been simulated using only classical resources. However, Bell tests provide the unimpeachable means of certifying these devices.

Such a QRNG is dubbed Device Independent QRNG or DI-QRNG for short [17–22]. The interesting thing here is that such Quantum Random number generators can be certified, without any knowledge of the inner workings of the device. It is therefore solely through non-local correlations present in quantum states that such a certification becomes feasible. Device Independent QRNG that meets the requirements of the output being perfectly unpredictable to anyone and meets the stringent norm absolute privacy. For the case of QKDs, the successful performance of loop-hole-free Bell tests, provides the information theoretical assurance that QKD supplies perfectly random keys to which only the authenticated parties are privy but no one else is.

3. Bell's inequalities

The fundamental assumption that the properties of a physical entity are independent and prior to any measurement is called realism. The premise that all physical processes are subject to relativistic causality is called locality. When observations are made at two locations and the only way in which the information of a measurement and it's outcome in the first location can be made available to the second location prior to the measurement made there is through a superluminal signal, we refer to the locations as being located in space-like separated regions. Any theory which asserts that measurements made at space-like separated regions cannot influence outcomes in other regions is called a local theory. All theories where both these conditions of locality and realism are maintained to be valid, are called local realistic theories [23]. Quantum Mechanics is in good part patently non-local and does not uphold realism. We say to a "good part" because separable states in quantum mechanics do not exhibit this property, only entangled states do. While the experimental certification of any local realistic theory or quantum mechanics as the correct description of nature is logically impossible, the consistency of one or the other with observations is feasible within experimental errors. We may test for the compatibility of local or local-realistic theories with experimental facts by supplementing these theories addition assumptions that account for common causes or prior correlations on two systems that had interacted earlier but are now located in space-like separated regions. Such theories are called local hidden variable theories. In trying to establish the appropriateness of a theory, it is customary to look for theories with fewest assumptions and their explanatory power over a wide variety of phenomena. A single failure would of course render it unacceptable. It is within these restrictions that one would look for the contradistinction between a theory or of possible set of theories with others. In the present case, we are interested in the differences in the predictions made by local realistic theories and quantum mechanics.

John Bell [23] proved an extraordinary and significant theorem that imposes quantitative limits on the correlations allowed by local realistic theories. The central result here is that the correlations exhibited by maximally entangled states exceed these limits. Before venturing into Bell discovery, it is first necessary to appreciate that the tests proposed by Bell are not tests on the validity of quantum mechanics per se. Bell tests merely provide an upper bound on the level of correlations that can attained by any local realistic theory. In quantum mechanics, given the state of a multipartite system, the decomposition of such a composite state into product states of the subsystems is in general not possible. For example, there are bi-partite systems $|\varphi\rangle_{AB}$ that can in general be written as a convex combination of product of the states of the sub-systems: $|\psi\rangle_{AB} \neq |\varphi\rangle_A \otimes |\chi\rangle_B$. The states that can be so-written are called separable states [2]. States which are not separable are called entangled states [24]. There are bi-partite states called Bell states that are maximally entangled in that, they exhibit perfect non-local correlations or anti-correlations. The sub-systems for such states could either have been generated through a common process of they could have interacted directly or indirectly interacted in the past and may describe a physical property subject to some conservation law. Rather than giving original references we directed the reader to a review article [24] for further details. As an example of Bell states we consider, two photons prepared through the process of spontaneous parametric down conversion (SPDC) [25] that could be prepared in the following Bell states which describe photons entangled in their polarization degree of freedom:

$$|\psi^{\pm}\rangle = \frac{1}{\sqrt{2}}(|HH\rangle \pm |VV\rangle) \tag{1}$$

$$|\phi^{\pm}\rangle = \frac{1}{\sqrt{2}}(|HV\rangle \pm |VH\rangle) \tag{2}$$

Photons being "flying qubits" each of the two photons could travel to two parties separated by an arbitrary distance. When either of the parties makes a measurement in the V/H basis, both photons in the $|\psi^{\pm}\rangle$ state would be found in the horizontal or vertical polarization with equal probability and the states of polarization of the two photons would be perfectly correlated. In case the $|\phi^{\pm}\rangle$, a perfect anti-correlation would be the result. Local measurements performed at spacelike separated regions, ensure the no-signaling condition. The Bell states could very well have been prepared in Bell states in some other basis set such as:

$$|\psi^{\pm}\rangle = \frac{1}{\sqrt{2}}(|DD\rangle \pm |AA\rangle) \tag{3}$$

$$|\phi^{\pm}\rangle = \frac{1}{\sqrt{2}}(|DA\rangle \pm |AD\rangle) \tag{4}$$

and so forth. In any case, Alice the source of the qubit and Bob the receipient of the qubit could choose to measure their photon in the {H, V} basis or the {D, A} basis. If the Bell states have been prepared with {H, V} polarizations states, only when both Alice and Bob measure with identical basis sets would they end with perfectly correlated or anti-correlated outcomes else their outcomes are perfectly random with respect to each other since {H, V} and {D, A} basis sets used by Alice and Bob for their local measurements are mutually unbiased. The initial states could then be subject to loop-hole-free Bell test by using the two black boxes procured from the supplier. In the case of an entanglement-based implementation of QKD, so long as it is ensured that no classical information such as the measurement out-comes leaks-out from Alice or Bob, the device outputs are secure. The notion of non-locality may best be understood through a Gedankenexperiment popularized by Popescu and Röhrlich [26]. We will first introduce one version of this experiment. The mathematical treatment laid out here closely follows the treatment in [27]. Let Alice and Bob be two stations that are space-like separated. Let these two stations be provided with two black boxes. These boxes have inputs x and y respectively, where, $x, y \in \{0, 1\}$. Let these two black boxes be designed to produce outputs a, b such that $a, b \in \{-1, +1\}$. The **Figure 1**, illustrates this game.

Quite independent of any physical theory, we are free to impose restrictions on the outputs for various possible inputs. The statistical outcomes of such games that can be repeatedly played these boxes is best described through conditional and joint probabilities. We are interested in computing the joint probability the outputs take

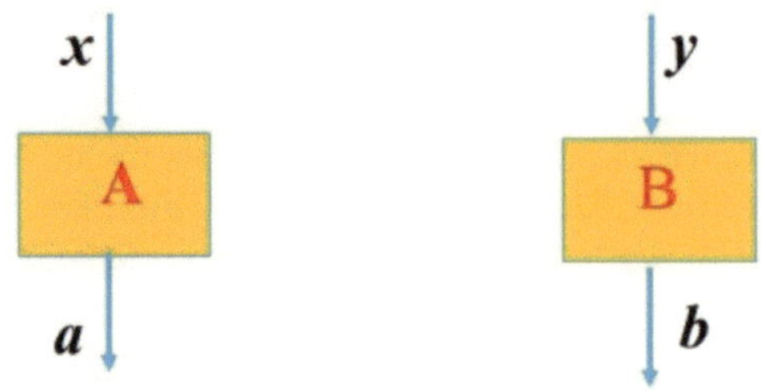

Figure 1.
A and B are two black boxes located in space-like separated regions. Inputs x, y to these boxes take value {0, 1}. The output of these boxes a, b assume values {−1, +1}. The joint probability $p(a,b|x,y)$ is the quantity of interest in this Gedankenexperiment.

conditioned by the input setting and their numerical values. Such a joint probability is written as $p(a,b|x,y)$. Here, we have simplified the notation by not writing the setting and the input values separately. When the two boxes are well-separated such that no signal traveling at a finite speed it generally expected that the outputs of the two boxes would be influenced only by the input settings and their values of each box and that the outputs would not be influenced by the input setting of the distant box. This assumption is called a *no-signaling* condition [1]. Under such a constraint, the joint probability would be written as:

$$p(a,b|x,y) = p(a|x).p(b|y). \tag{5}$$

In writing the joint probability in terms of the product of probabilities as above, the additional assumption is that there are no common causes or past influences that would bring about a correlation between the two boxes. To account for all such possibilities, we may rewrite the conditional probability above as:

$$p(a,b|x,y,\lambda) = p(a|x,\lambda).p(b|y,\lambda) \tag{6}$$

where λ accounts for all possible common causes and influences. The parameters (s) are sometimes referred to as hidden variables. As an illustrative example of such factors, let us consider two individuals at two distant location who go shopping for a soap dish or a toothbrush and that these two objects generally come in blue or orange color. If a common supplier had supplied a stock of only blue soap dishes and orange toothbrushes, then it should come as no surprises that whenever the two individuals buy the same object they end-up with the same color and whenever they buy different objects, they are of a different color. To factor-in such possibilities, we may rewrite the above joint conditional probability in terms of the product of the individual conditional probabilities. When the Joint conditional probability can be factored as above, we refer to the condition as being non-local. By assuming that the variable λ has a well-defined probability distribution function $\sigma(\lambda)$ that does depend on the input settings i_a, i_b of either of the boxes, we can integrate over that it might take during various runs of the experiment and arrive at

$$p(a,b|x,y) = \int d\lambda\sigma(\lambda)p(a|x,\lambda).p(b|y,\lambda) \tag{7}$$

This condition is then a formal statement of the locality condition. This gist of this statement is that any local operations carried out on either of the two stations oughtn't have any influence on the other station, when the two stations are in space-like separated regions. It is implicitly assumed that the choice of input setting is independent of $\sigma(\lambda)$ which is itself independent of the input settings. In actual implementations, the input settings are chosen with the aid of quantum random number generators. Whenever

$$p(x,y|a,b) \neq p(x|a).p(y|b) \tag{8}$$

the two events are not independent of each other but are correlated. In Bell's original formulation [bell24], he considered only perfect correlations or anti-correlation in the outputs. The CHSH inequality considers the experimentally realistic situation and based on the computation of expectation values of the outputs. Given $x,y \in \{0,1\} \wedge a,b \in \{-1,+1\}$, the expectation value or the average value over an ensemble of repeated measurement of identically prepared states is given by:

$$\langle a_x b_y \rangle = \sum_{a,b} ab\, p(ab|xy) \tag{9}$$

Under conditions of objective locality or local realism, the following equality holds:

$$S = \langle a_0 b_0 \rangle + \langle a_0 b_1 \rangle + \langle a_1 b_0 \rangle - \langle a_1 b_1 \rangle \leq 2 \tag{10}$$

With quantum systems, S could exceed this value because non-separable states are of a significantly different nature compared local realistic theories. To appreciate this, we may write Bell states in terms of a computational basis as for instance

$$|\psi\rangle = \frac{1}{\sqrt{2}}(|00\rangle \pm |11\rangle) \tag{11}$$

The vectors $0\rangle \wedge 1\rangle$ are the eigen vectors of the Pauli operator σ_z.Identifying the inputs x and y with measurements along vectorial directions x and y respectively, the quantum mechanical expectation value $\langle a_x b_y \rangle - x.y$. When a mutually unbiased basis (MUBS) [28] choice is made for x, y, it is trivial to show that $S \leq 2\sqrt{2}$.

4. Experimental tests of Bell's inequalities

Let us now consider the following experiment wherein there are two experimental stations as discussed in the earlier Gedankenexperiment but with a small twist: Here, S is a SPDC source emitting a pair of polarization entangled photons in one of the Bell states and let a and b be randomly obtained from certified QRNGs located and securely isolated at the stations **a** and **b** respectively. The two inputs of are then used to choose between the two mutually unbiased $\{V,H\}$ and $\{D,A\}$ where, V/H refers to the vertical/horizontal basis and D/A refers to diagonal/anti-diagonal basis sets respectively (**Figure 2**).

The state emitted from the source is one of the Bell states and let us assume without loss of generality, that the state is $\psi^+\rangle$ Spontaneous parametric downconversion is a probabilistic process with a very low probability of emitting an entangled photon pair. Hence, for optimal pump laser powers, the probability of multiple pairs being emitted is extremely low. After considering the travel time and setting a coincidence window, when both detectors detect photons, it most likely that the pair of photons were emitted simultaneously and are in an entangled state. The source may be suitably characterized through Quantum State Tomography

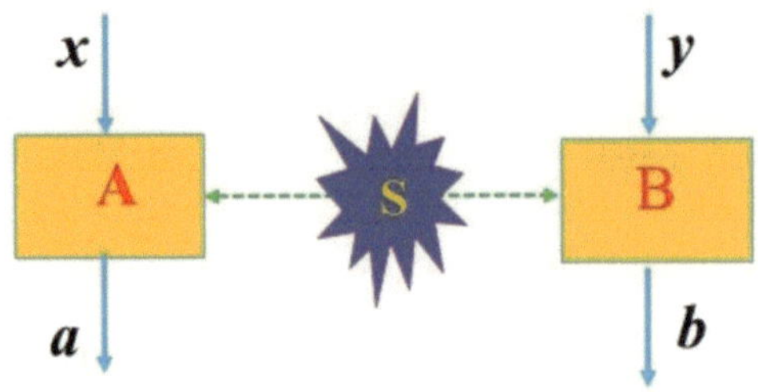

Figure 2.
S is a entangled photon source emitting photon in a Bell state with one photon of each pair reaching stations A and B located in space-like separated regions. Inputs x, y to these boxes take value {0, 1}. The basis choice at either station is made based on the random inputs x, y. The output of these boxes are a, b assuming values {−1, +1} as earlier.

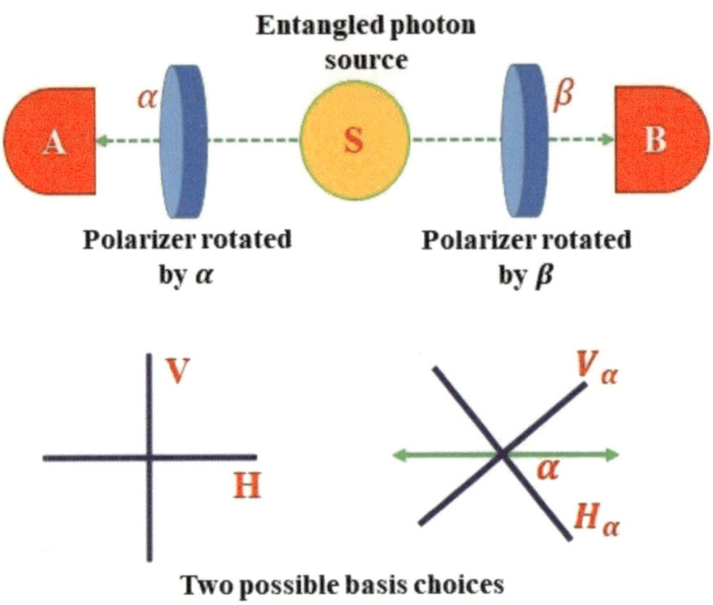

Figure 3.
Basis choices for measurements may be made independently at either station such as H/V, H_α/V_α for arbitrary α.

(QST) [3]. Usually, local corrections of polarization may be corrected through suitable polarization controllers located at A and B. Once the steps are done and the source is well-characterized, projective measurements are carried out at each of the stations. To carry out measurements, the basis choice at each station is carried out rotating the polarizers by some angle. We may however choose to measure in rotated basis as illustrated (**Figure 3**).

The rotated basis vectors may be expressed as:

$$|H_\alpha\rangle = \cos\alpha|H\rangle - sin\alpha|V\rangle, |V_\alpha\rangle = \sin\alpha|H\rangle + cos\alpha|V\rangle \qquad (12)$$

If the polarizers at A and B are rotated by angles α and β respectively and an ensemble of measurements are carried out on identically prepared states, quantum mechanics predicts the probability of obtaining coincidence counts when the vertical polarization is measured to be:

$$P_{VV} = \left|\langle V_\alpha V_\beta|\psi^+\rangle\right|^2 = \frac{1}{2}\cos^2(\beta - \alpha) \tag{13}$$

and likewise,

$$P_{HH} = \left|\langle H_\alpha H_\beta|\psi^+\rangle\right|^2 = \frac{1}{2}\cos^2(\beta - \alpha) \tag{14}$$

$$P_{VH} = \left|\langle V_\alpha H_\beta|\psi^+\rangle\right|^2 = \frac{1}{2}\sin^2(\beta - \alpha) \tag{15}$$

$$P_{HV} = \left|\langle H_\alpha V_\beta|\psi^+\rangle\right|^2 = \frac{1}{2}\sin^2(\beta - \alpha) \tag{16}$$

Defining:

$$E(\alpha, \beta) = P_{HH} + P_{VV} - P_{VH} - P_{HV} \tag{17}$$

and

$$S = \left|E(a,b) - E(a,b')\right| + \left|E(a',b) - E(a',b')\right| \tag{18}$$

For certain angles of the polarizers, this parameter S can acquire values greater than 2. For instance, for $a = \frac{\pi}{4}, a' = 0, b = \frac{-\pi}{8}, b' = \frac{\pi}{8}$, the value of $S = 2\sqrt{2}$. Carefully performed measurements on any of the Bell states are in good agreement with the quantum mechanical predictions This number can be easily shown to be ≤ 2 for any arbitrary local realistic theory [29]. This inequality is called the CHSH inequality. Thus, a value of S exceeding 2 is indicative of the presence of non-local correlations.

5. Loop-hole-free Bell tests

The actual measurement of quantum states to check for violation of CHSH inequalities involves the use of devices that involves losses and detectors that have an efficiency μ much less than 1. In such a case, the CHSH inequality is obtained by evaluating the expectation values conditional to coincident counts in both the detectors. This is necessary because of finite losses in the communication channels and μ being less than 1 [].

$$S = |E(a,b)|coin) - E(a,b')|coin)| + |E(a',b)|coin) + E(a',b')|coin)| \leq \frac{4}{\mu} - 1 \tag{19}$$

Therefore, S > 2 if and only if $\mu > 0.828$ and hence, Bell's inequality violation would be seen only if the detector efficiency is better than this value. Superconducting nanowire single photon detectors are nowadays commercially available. When detectors of this efficiency are not available, substantial number of coincidences indicating the presence of entangled pairs of photons go undetected. Under these conditions, the sub-ensemble of coincidence detected are assumed to truly representative of the statistics of the entangled photon pairs emitted by the source. This assumption is called the fair sampling assumption. If this assumption holds, then $S \leq 2$ for all local-realistic theories. There is in fact yet another assumption that pertains to detectors which would result in false positive entangled pair detections. Because of experimental expediency, a coincidence event is defined as pairs detected within a coincidence pre-assigned time window. Even uncorrelated pairs of events could result in a seeming coincident event when one or the other photon is delayed by a suitable time interval. Such an occurrence could result in the Bell's inequality violation for classical source.

In an actual quantum key distribution protocol, it is assumed that the choice of the measurement basis is made randomly and independently. The actual detection set-up looks like that indicated in the **Figure 4** below when either of the following measurement basis choices $\{H/V\}$ or $\{D/A\}$ is made for polarization entangled photon at the source. With such an arrangement, the non-polarizing 50:50 beam splitter sends the incoming photon to either of the polarizing beam splitters with equal probability and hence a random basis choice $\{V, H\}$ or $\{D, A\}$ is made. This choice is not pre-determined and is perfectly random in nature.

Other than the loopholes mentioned earlier, there are very many other possible loopholes that could vitiate experimental demonstration of Bell's inequality violation. For example, the memory loophole wherein it may be posited that somehow the experimental apparatus retains the details of the previous measurements, thereby rendering the conclusions questionable. Another significant loophole is called the locality loophole under a presumed superluminal communication between the two stations. There are many other possible loopholes and remedial measures that could be taken to close them. We shall desist from going into each of them. The interested reader may refer to [1] and some references contained therein. Suffice to say that is experimentally demanding to demonstrate that all the loopholes have been closed in a single experimental. However, closing one or more loophole but not all have been demonstrated in numerous experiments. There are a couple of experiments which claim to have succeeded in closing all the loopholes. The possibility of someone coming up with an ingenious loophole proposal, however improbable, cannot be ruled out (**Figure 4**).

The watertight loop-hole-free experimental demonstration of information security requires a throughgoing analysis of the complete experimental conditions as well as characterization of components used in the experimental apparatus for deviation from the idealized system and a careful characterization of their imperfection. Alternately, the experimental apparatus is accepted as being unreliable. In the latter case, one is left with the option of having rely on a careful statistical

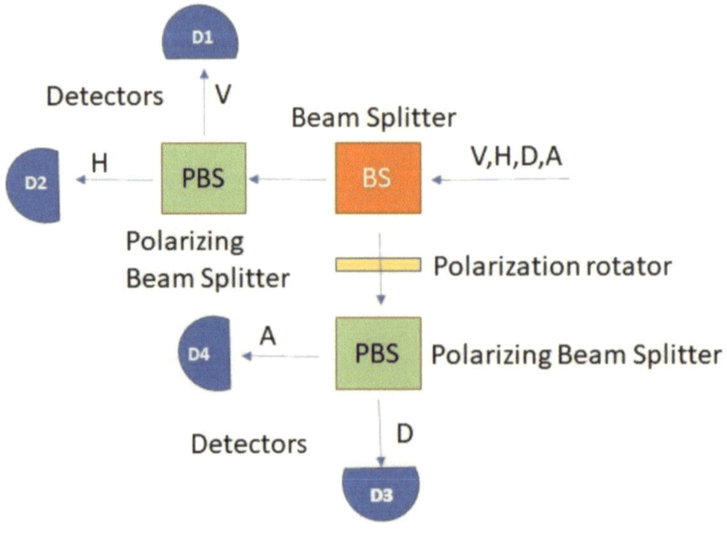

Figure 4.
The experimental arrangement for choosing randomly between two possible polarization choices for polarization is indicated.

analysis of observed nonlocal correlations of the data coupled with an experimental apparatus that comes close to being loophole free to the extent that is possible.

6. Device independence

Quantum Random generators and QKD system are devices that use many components that are imperfect, and their behavior deviates from the ideal systems assumed in theoretical models and as we had argued earlier, their complete characterization is well-nigh impossible because the physics of every single component that is used to build these systems needs to be modeled to perfection.

There exists however an alternate possibility, to obtain certified QRNGs and QKD devices which is inaccessible to classical tools. Surprisingly, in the absence of any superluminal communication, it is possible to use experimentally observed non-local correlation for this purpose. Such an approach does not require the modeling of the devices in question. What is more, the devices can be used as black boxes which have been supplied by a completely untrusted source. In the case of QRNG, the device needs to be intrinsically random, and the privacy of the random sequences needs to be guaranteed. In other words, the QRNG should be PPRNG. The use of non-local correlations certified through loophole-free Bell tests should drive us to physical limits of privacy of the random key generated by Alice and Bob. We shall treat QRNGs and QKDs separately and look at the overall outlook of device independent approaches.

6.1 Device independence QRNG (Di-QRNG)

The very definition of randomness is fraught with problems of philosophical nature. We had earlier alluded to differences between pseudo-random number generators (PRNGs) realized through algorithmic techniques, true random number generators (TRNGs) of epistemic origins and quantum random generators (QRNGs) which is believed to ontological in nature. The task at hand is to certify that the device at hand is a genuine QRNG. The output of such a device should be certifiably random not only to the user but every possible user. The density matrix describing N perfectly random output of 0 or 1 with equal probability is described by completely mixed density matrix given in the computational basis by $\hat{\rho} = \frac{1}{2}$. When this output is perfectly isolated from the environment is described by the product state:

$$\hat{\rho} \bigotimes \hat{\rho}_E \tag{20}$$

Where, $\hat{\rho_E}$ is the state of environment [2]. Since the nature of random sequences generated is of a physical origin, perfect and perfectly private randomness should be certifiable through quantum process. Therefore, nonlocal correlations witnessed by Bell's inequality violations could be employed to certify the QRNG. It stems from the fact that Bell tests on entangled sources generate perfectly random sequences under local measurements. The perfect randomness of local measurement outcomes attests to the fact that such measurements have been made on maximally entangled pure states. Maximally entangled states are subject to monogamy conditions [2] and hence cannot be entangled with environment. The correlations between measured outcomes are presented in terms of conditional probabilities as explained in the earlier sections. The catch however is that the demonstration of Bell's inequality violation should be loop-hole-free! The worst-case scenario for an unreliable QRNG is when the supplier of the device has packaged the device with pre-generated random numbers. Such numbers would pass all tests of randomness but would hardly be private, since the supplier could have made copies of the same. The basic idea behind Quantum Mechanics certified randomness is that Bell's inequality violations can guarantee that the observed randomness is not pre-generated. Two conditions need to be fulfilled for demonstrating device independence and they are 1. The basis choice (a.k.a the measurement setting) in the two stations in Bell tests are independent of experimental devices and of any prior information of each of them as might be available and 2. The measurement outcomes of each station are independent of the measurement setting in the other station. The "Free-will" choice is an assumption that is ill-proven and anthropomorphic. In engineered system free-will is replaced by a source of intrinsic private randomness. This is rather curious because, the entire exercise that is undertaken for DI has to do with the certification of such sources. The second condition is however readily satisfied so long as the stations cannot communicate with each other (no signaling condition). This step could involve some public source of quantum random numbers. The initial seed could also be enlarged through the process of random number expansion see [random num exp] and references contained therein. The basic idea is that the numbers obtained through a Bell test are a source of certified randomness. It may be noted here that at least two devices are required to test for device independence.

In summary, DI-QRNGs [30, 31] use Bell inequality violations to certify the quantum state generated within the devices are pure entangled states. The purity of the quantum state ensures an absence of correlation not only between the devices at stations A and but also with the environment and observers. Under a local measurement of the sub-system of a pure entangled state generates a completely mixed states resulting in perfect randomness of the output as certified by some entropic measure. Bell certified randomness is of a quantum nature as classical devices always do not violate Bell inequalities. Many DI-QRNG proposals as well experimental realization by various types are available in the literature. We will not attempt any systematic review of the literature. Quantum random number generators which rely on non-locality testified by Bell tests are also called self-testing QRNGs [17], the main problem with such devices is that they are presently too slow.

6.2 Device independence QKD (DI-QKD)

The one-time pad is a provably secure method of encryption [32]. The principle behind one-time pad is extremely simple: To encrypt a message bitstring of N bits called the plain text, a random bitstring of the same size called the key is generated. Then a modular addition of the key and plain text is carried out to create a bitstring called the ciphertext. The ciphertext is then communicated through a public channel to the recipient with whom the key is shared through a secure means. A modular addition of the ciphertext with key by the recipient, yields the plain text or message. Finding the means of sharing the key between the sender and the recipient of the message is called the key distribution problem. Traditionally, a trusted courier was given this job. This of course is not a viable option for encrypting terabits of data per second in the modern context.

A QKD system is device that acts a trusted courier of key between two parties. The security of such systems by the rules of quantum mechanics. The carriers of information are photons derived from a weak coherent source (attenuated laser pulses) of entangled photon sources. The quantum state of a single photon cannot be copied perfectly (No-cloning theorem) and a quantum state will be disturbed by the act of observation due to the Heisenberg Uncertainty principle. These quantum features of photons are exploited to ensure provable security of the key that is exchanged between two parties. Typically, the sender prepares the photons by choosing randomly different bases for measurement and communicates each photon in one or the other eigenstates of the bases. The eigenstate is again chosen randomly. Usually, the sender uses a QRNG for this purpose. Likewise, the receiver chooses to measure the photon in one or the other basis. After exchanging a large number of photons, the basis choice made by both parties are compared and only those cases where the choice is the same, the corresponding measured outcomes are retained. Under ideal circumstances, this process would result in a privately shared keys that are identical. Practical Quantum Key distributions whether implemented on optical fibers or free-space are however inherently noisy because of photonic losses, and changes in the state during transmission. Such devices also use sources of single, heralded or entangled photons that are not perfect and detectors that usually have efficiencies below the requisite efficiency of $\sim$83%. These devices also use a variety of commercial components that are prone to side-channel attacks and are not the ideal ones used in a theory. Thus, the claim of provable security does not apply for practical systems. This makes QKD devices vulnerable to a variety of side-channel attacks. Thus, the raw keys obtained through the quantum channel have to subjected a series of post-processing steps for the generation of the final keys. Since most of side-channel attacks were on the detector side, measurement device independent QKDs were proposed and implement. The final frontier of physical limits of privacy can be guaranteed only by device independent QKD systems. As in DI-QRNGs, DIQKD [33–35] also necessitate the performance of Bell tests between two distant parties. Bell test typically use the Clauser-Horne-Shimony-Holt (CHSH) variant of Bell tests, which employs maximally entangled states. The rate of key generation, distance of transmission and security assurance levels are all inter-related in practical systems. Usually when low efficiency detectors are employed and significant line losses occur, fair sampling is implicitly assumed. In DI-QKD or measurement device independent MDI-QKD [36–38], the measuring device is with the quantum hacker Eve and fair sampling arguments are no longer valid. Security of DI-QKD depend on the monogamy of shared correlations between maximally entangled photonic states. As in the case of DI-QRNG, device independence accrues through the conduct of loop-hole-free Bell tests. Mayers and Yao [33] proposed an early version of DI-QKD dealing with specific case of imperfect sources. In this pioneering work, they proposed that the security of a QKD protocol may be tested using entanglement-based protocols. Jonathan Barrett, Lucien Hardy, and Adrian Kent showed that single shared bit with guaranteed security can be exchanged though the use Bell tests. Since these early results a variety of proposal and proof of concept implementations have been published in the literature. As in the case of QRNGs, DI-QKD systems are extremely difficult to implement because, the ultimate guarantee of physically assured privacy relies on the performance of loop-hole free Bell tests.

7. Conclusions

In this chapter, we have attempted to provide a brief overview of device independent QRNGs and QKD systems that exploit Bell test to guarantee privacy and randomness. In a reasonably complete manner, no attempt has however been made to review the field in a systematic and cogent fashion. The realization of device independence based on Bell's inequality violation was discussed. The central idea is to show that device independence of quantum devices is as hard to achieve as loop-hole-free Bell tests. The performance of such tests requires random generators that are provably secure. While there are large number of reports in the literature where subsets of possible certain loopholes have been closed in certain experiments, there are but a couple of them that claim to have closed all loopholes.

Acknowledgements

I would like to thank Srimathi and Vakul for being there for me. I thank Dr. Srinivasan, DIAT, for his help with the figures. I thank DIAT for providing a very congenial academic environment and the support it provides. I wish to convey my sincere thanks and gratitude for the patience and diligence of Maja Bozicenic, Author Services Manager and the Editors.

Author details

Gopalan Raghavan
School of Quantum Technology, Defence Institute of Advanced Technology, Pune, India

*Address all correspondence to: go.raghavan@gmail.com

References

[1] Barrett, J., Hardy, L., and Kent, A. No-signalling and quantum key distribution. Phys. Rev. Lett. 95, 010503 (2005).

[2] Masanes, L., Acin, and Gisin, N. General properties of nonsignaling theories. Phys. Rev. A. 73, 012112 (2006).

[3] Michael A. Nielsen and Isaac L. Chuang. Quantum Computation and Quantum Information. Cambridge University Press.

[4] Ekert, A. Phys. Rev. Lett. 67, 661-663, 1, 991 (2005).

[5] Rarity, J., Owens, P., and Tapster, P. Quantum random-number generation and key sharing. J. Mod. Opt. 41, 2435–2444 (1994).

[6] Liu, Y. et al. Device-independent quantum random-number generation. Nature 562, 548–551 (2018).

[7] Xiongfeng Ma, Xiao Yuan, Zhu Cao, Bing Qi, and Zhen Zhang. Quantum random number generation, npj Quantum Information (2016) 2

[8] The Marsaglia Random Number CDROM including the Diehard Battery of Tests of Randomness. Available at (http://www.stat.fsu.edu/pub/diehard/) (2008).

[9] Rukhin, A. et al. A Statistical Test Suite for Random and Pseudorandom Number Generators for Cryptographic Applications (Special Publication 800–22 Revision 1, National Institute of Standards and Technology, 2008); available at (http://csrc.Nist.gov/publications/PubsSPs.html).

[10] Bell's inequality, N. Brunner, D. Cavalcanti, S. Pironio, V. Scarani, and S. Wehner, Rev. Mod. Phys. 86, 419 (2014)

[11] Hensen, B. et al. Loophole-free bell inequality violation using electron spins separated by 1.3 kilometres. Nature 526, 682–686 (2015).

[12] Shalm, L. K. et al. Strong loophole-free test of local realism*. Phys. Rev. Lett. 115, 250402 (2015).

[13] Giustina, M. et al. Significant-loophole-free test of Bell's theorem with entangled photons. Phys. Rev. Lett. 115, 250401 (2015).

[14] Einstein, A., Podolsky, B. and Rosen, N. Can Quantum-Mechanical Description of Physical Reality Be Considered Complete? Phys. Rev. 48, 777-780 (1935)

[15] N. Bohr. Can Quantum-Mechanical Description of Physical Reality Be Considered Complete? Phys. Rev. 48, 696 (1935)

[16] Uchida, A. et al. Fast physical random bit generation with chaotic semiconductor lasers. Nat. Photon. 2, 728–732 (2008).

[17] Lunghi, T. et al. Self-testing quantum random number generator. Phys. Rev. Lett. 114, 150501 (2015).

[18] Cao, Z., Zhou, H. and Ma, X. Loss-tolerant measurement-device-independent quantum random number generation. New J. Phys. 17, 125011 (2015).

[19] Liu, Y. et al. Device-independent quantum random-number generation. Nature 562, 548–551 (2018).

[20] Marangon, D. G., Vallone, G. and Villoresi, P. Source-device-independent ultrafast quantum random number generation. Phys. Rev. Lett. 118, 060503 (2017).

[21] Acín, A., Massar, S. and Pironio, S. Randomness versus nonlocality and

entanglement. Phys. Rev. Lett. 108, 100402 (2012).

[22] Avesani, M., Marangon, D. G., Vallone, G. and Villoresi, P. Source-device-independent heterodyne-based quantum random number generator at 17 gbps. Nat. Commun. 9, 5365 (2018).

[23] J.S. Bell, Physics I, 195, 1965.

[24] Ryszard Horodecki, Pawel Horodecki, Michal Horodecki, Karol Horodecki. Quantum entanglement. Rev. Mod. Phys, 81, 865 (2009)

[25] P. Kwiat; et al. (1995). New High-Intensity Source of Polarization-Entangled Photon Pairs. Phys. Rev. Lett. 75 (24): 4337–4341.

[26] Popescu, S. and Rohrlich, D. Quantum nonlocality as an axiom. Found. Phys. 24(3) 379-385 (1994)

[27] Antonio Acin and Luis Masanes. Certified Randomness in Quantum Physics. Nature 540, 213 (2016)

[28] Benston Ingemar (2007). Three Ways to Look at Mutually Unbiased Bases. AIP Conference Proceedings. 889. pp. 40–51

[29] Clauser, J. F., Horne, M. A., Shimony, A. and Holt, R. A. Proposed experiment to test local hidden-variable theories. Phys. Rev. Lett. 23, 880–884 (1969).

[30] Giustina, M. et al. Bell violation using entangled photons without the fair-sampling assumption. Nature 497, 227 (2013).

[31] Pironio, S. et al., Random Numbers certified by Bell's Theorem. Nature 464, 1021-1024 (2010).

[32] Vernam, G. S. Cipher printing telegraph systems for secret wire and radio telegraphic communications. J. Am. Inst. Electr. Eng. 45, 109–115 (1926).

[33] Mayers, D. and Yao, A. Self-testing quantum apparatus. Quant. Inform. Comput. 4 273-286 (2004).

[34] Acin, A., Gisin, N. and Masanes, L. From Bell's Theorem to Secure Quantum Key Distribution. Phys. Rev. Lett. 97, 120405 (2006).

[35] Masanes, L., Pironio, S. and Acin, A. Secure device-independent quantum key distribution with causally independent measurement devices. Nature Comm. 2, 238 (2011).

[36] Liu, Y. et al. Experimental measurement-device-independent quantum key distribution. Phys. Rev. Lett. 111, 130502 (2013)

[37] Tang, Z. et al. Experimental demonstration of polarization encoding measurement-device-independent quantum key distribution. Phys. Rev. Lett. 112, 190503 (2014).

[38] Yin, H.-L. et al. Measurement-device-independent quantum key distribution over a 404 km optical fiber. Phys. Rev. Lett. 117, 190501 (2016).

Chapter 8

Quantum Information Science in High Energy Physics

Oliver Keith Baker

Abstract

We demonstrate that several anomalies seen in data from high energy physics experiments have their origin in quantum entanglement, and quantum information science more generally. A few examples are provided that help clarify this proposition. Our research clearly shows that there is a thermal behavior in particle kinematics from high energy collisions at both collider and fixed target experiments that can be attributed to quantum entanglement and entanglement entropy. And in those cases where no quantum entanglement is expected, the thermal component in the kinematics is absent, in agreement with our hypothesis. We show evidence that these phenomena are interaction independent, but process dependent, using results from proton-proton scattering at the Large Hadron Collider (LHC) and antineutrino-nucleus scattering at Fermilab. That is, this thermal behavior due to quantum entanglement is shown to exist in both the strong and electroweak interactions. However, the process itself must include quantum entanglement in the corresponding wave functions of interacting systems in order for there to be thermalization.

Keywords: Quantum Entanglement, Entanglement Entropy, High Energy Physics

1. Introduction

A complete understanding of multi-particle production dynamics continues to be a challenge for theory in high energy collisions. The full description of real-time dynamical evolution in a strongly coupled non-Abelian gauge theory can be notoriously difficult. The availability of large, diverse, and high quality accumulated proton-proton (*pp*) and heavy ion (HI) collision data from the energy frontier at the Large Hadron Collider (LHC) and at Brookhaven National Laboratory's Relativistic Heavy Ion Collider (RHIC) are providing new insights into puzzling behavior observed in these strong interaction processes. Additionally, studies of similar intensity frontier anomalies in (anti)neutrino-nucleus scattering gathered recently also benefit from newly accumulated event statistics at Fermilab. The transverse momentum distribution of differential cross sections in *pp* collisions shows process-dependent behavior that requires a more subtle explanation compared to previous ideas.

For example, the differential distribution of charged hadrons resulting from *pp* collisions are observed to exhibit both a well-understood quark and gluon "hard scattering" component at high transverse momenta that can best be described by a power-law fit to the distribution, and a component at low transverse momenta that exhibits a less well-understood "thermal" behavior that is best represented by an exponential fit to the data [1, 2]. See [3, 4] for a review. The sum of these two contributions is necessary to properly characterize the transverse momentum differential distribution. And yet, for diffractive events where there is a large rapidity gap in the transverse momentum distribution of the differential cross section for photo-produced muon pairs, there is only a hard-scattering contribution; the thermal component disappears in this class of processes even though these inelastic events produce large number of hadrons. Charged-current weak interactions also exhibit differential cross section momentum distributions that have process-dependent results. As in the case of strong interactions, for anti-neutrino scattering from a nucleon inside a nucleus, there is also a "hard scattering" component at high particle momenta that is manifested by a power-law fit to the momentum distribution of the differential cross section. And at low momenta, the differential cross section behavior is best described by an exponential fit to the momentum distribution in this region. Furthermore, in coherent anti-neutrino scattering from a nucleus, where the nucleus remains intact, there is no thermal component to the differential cross section momentum distribution, only the hard-scattering component [5].

It is interesting to consider the possibility that these two interactions of vastly differing collision or scattering energies are different manifestations of a single underlying fundamental process. There is growing interest in the link between quantum entanglement, entanglement entropy (EE), and high energy physics presently. Here, we describe a relationship between quantum entanglement in the nucleon wave functions associated with the hadron collisions at the LHC experiments and with the electroweak scattering in Fermilab experiments. There are several examples of these relationships in theoretical physics: Research on the dynamics of quantum entanglement and entanglement entropy in the regime of small Bjorken-x in deep inelastic scattering, in electromagnetic interactions [6]. Additionally in [1] the case is made that quantum entanglement between partons inside a nucleon can be probed by deep inelastic lepton scattering. Deeper insights into these dynamics involving entanglement entropy in the regions of black holes is now provided by the AdS/CFT correspondence [7]. Well-understood quark-antiquark correlations are now improved even more due to considerations of entanglement entropy in Lattice Gauge Theory [8] that are or soon will be augmented by new and planned parton distribution functions data in particle and nuclear physics. Newer intuition into this relationship that involves quantum entanglement, entanglement entropy, thermal behavior, is gained from heavy ion and proton-proton collisions [9, 10], nuclear shadowing effects [11], and chiral symmetry breaking [12].

In [1, 2, 13], it is proposed that the thermal component is a result of entanglement between causally disconnected parts of the nucleon in the interaction. For this reason inelastic pp collisions exhibit a thermal component, while diffractive collisions, where the nucleon as a whole is probed, give rise to only the hard-scattering component, and no thermal behavior. That the thermal component is a consequence of quantum entanglement between different regions of the colliding protons wave functions in pp colisions is proposed in [14]. A thermal behavior can, in the collisions where there are large number of particle produced in the final state, be the result of rescattering among the produced final state particles. However, the thermal component was shown to not only be present in the transverse momentum distribution of charged hadrons, but also in the transverse momentum spectrum of Higgs bosons production and decay differential cross section, resulting from the collisions. For Higgs boson production in pp collisions, there are very few final state particles for rescattering compared to heavy ion collisions for example.

If quantum entanglement and EE are responsible for the thermal behavior in charged hadron as well as Higgs boson production in pp collisions at the LHC, then it should also be observed in neutrino-nucleus scattering, where only a fraction of the struck nucleon in the interaction is probed by the exchanged charged-current probe. As explained in [5] the observation of a thermal component exists in the momentum distribution of neutral pions from antineutrino-nucleon scattering, while no such component is present in coherent antineutrino-nucleon scattering. This absence of a thermal component in the latter case is due to the fact that antineutrinos probe the nucleus as a whole in coherent scattering; there is no un-probed region of the nucleus which can be entangled with the probed region. So with no quantum entanglement in the interaction, the thermal component in the momentum distribution of the differential cross section is absent, as expected.

These topics are presented and discussed in this chapter. A brief description of the theory motivating this proposed link between quantum entanglement, entanglement entropy, and thermal behavior in pp collisions is given in Section 2 and subSection 2.1. This is followed by experimental results of transverse momentum distributions in pp collisions at 13 TeV collisions energy. Charged hadron production where (and why) both hard scattering and thermal components are present in the differential distribution are described in subSection 2.2. The absence of the thermal component in diffractive production of muon pairs in the reaction $pp \to \mu^+\mu^- X$ is explained in subSection 2.3. The interesting need for superposition of both the hard scattering and thermal components to describe the transverse momentum distribution of the Higgs boson is presented in subSection 2.4. Section 3 includes the presentation of this phenomena in charged current weak interactions. Concluding remarks are given in Section 4.

2. Entanglement entropy and thermal behavior in the strong interaction

We begin by considering the possibility that the observed thermalization in pp collisions is the result of a sudden perturbation or rapid "quench" due to the high degree of entanglement inside the protons involved in the collision [15]. The link between quantum thermalization and quantum entanglement is shown to exist in an experimental quench in Bose-Einstein condensates of Rb atoms in atomic and condensed matter physics [16]; the rapid eigenstate thermalization was found to be the result of a quantum entanglement. In pp process described here, low momenta correspond to late times after the collision. The thermal behavior begins to dominate over the hard scattering component in the transverse momentum distribution at late times. This is consistent with theoretical studies in $(1+1)$-dimensional conformal field theories of quenches in entangled quantum systems [17–19] where a system can be described by a generalized thermal Gibbs ensemble at late times.

Since a high-energy collision can be viewed as a rapid quench of the entangled partonic state [15], it is thus possible that the effective temperature inferred from the transverse momentum distributions of the secondaries in a collision can depend upon the momentum transfer, that is an ultraviolet cutoff on the quantum modes resolved by the collision. In analyzing the high-energy collisions with different characteristic momentum transfer Q we thus expect to find different effective temperatures $T \sim Q$. We can also look at the inelastic events characterized by a rapidity gap, where the proton is probed as a whole, and no entanglement entropy arises [15]. In this case, if the quantum entanglement is responsible for the thermalization, we expect no thermal radiation.

The presence of both a thermal and a hard scattering component in inclusive deep-inelastic scattering at HERA has been observed [20]. And the absence of this thermal component in processes characterized by a rapidity gap is also manifested in these studes. In diffractive events where there is a rapidity gap, the entire proton wave function is involved in the scattering process. In diffractive scattering, the proton remains fully intact in the central part of the collider detector where scattering takes place. There is no entanglement entropy due to different regions of the proton wave function being involved in the scattering in different ways. This observation points to a connection between this thermalization and quantum entanglement between different parts of the proton wave function. This link is described in the next section.

2.1 Entanglement and thermalization in high energy collisions - theory

A brief summary of these proposition is as follows. The hard process in *pp* collisions probes only the part of the proton wave function that is localized in a region of space denoted here as **A**. For a hard process such as the one shown in **Figure 1** this region has a transverse size that is less that the full proton diameter and, in the proton's rest frame, longitudinal size in terms of Bjorken-x is $\sim (mx)^{-1}$, where m is the proton mass.

In this same figure, the spatial region **B** is complementary to **A**, that is, the entire space is **A**∪**B**. Hard processes have their origin in the physical states inside the region **A**. These are states in a Hilbert space $\mathcal{H}_A$ of dimension n_A. Unobserved states (not part of hard scattering) in the region **B** belong to the Hilbert space $\mathcal{H}_B$ of dimension n_B. With this picture, the protons prior to the collision, both composite systems in **A**∪**B** (the entire proton in each case) are then separately described by the vector represented as, for example, $|\psi_{AB}\rangle$ in a tensor product of the two spaces $\mathcal{H}_A \otimes \mathcal{H}_B$:

$$|\psi_{AB}\rangle = \sum_{i,j} c_{ij} \; |\varphi_i^A\rangle \otimes |\varphi_j^B\rangle, \tag{1}$$

where c_{ij} are the elements of the matrix C that has a dimension $n_A \times n_B$. In the case where there are states $|\varphi^A\rangle$ and $|\varphi^B\rangle$ that $|\psi_{AB}\rangle = |\varphi^A\rangle \otimes |\varphi^B\rangle$, where that the sum (Eq. (1)) contains only one term, then the state $|\psi_{AB}\rangle$ is product state that is separable. In the case where it is not separable, $|\psi_{AB}\rangle$ is entangled.

$|\psi_{AB}\rangle$ is called a bi-partite system that, making use of the Schmidt decomposition theorem, can be expanded as a *single* sum in n instead of a double sum over ij.

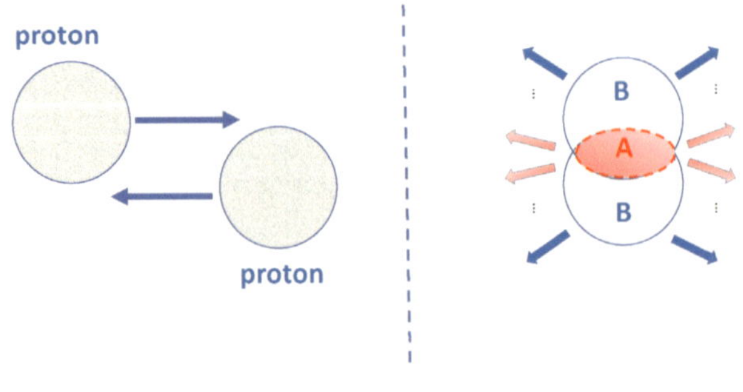

Figure 1.
*Characterization of the entanglement entropy in pp collisions. In the leftmost depiction, the collider protons, before collision, are both pure states. In the rightmost depiction, during the pp scattering, there exists the proton overlap collision region (**A**) and the overlap spectator region (**B**).*

$$|\psi_{AB}\rangle = \sum_n \alpha_n |\psi_n^A\rangle|\psi_n^B\rangle \tag{2}$$

Here $|\psi_n^A\rangle$ and $|\psi_n^B\rangle$ are orthonormal sets of states (properly chosen) localized in the domains **A** and **B**, respectively. And α_n are real, positive numbers that are the square roots of the eigenvalues of matrix $CC^\dagger$.

The density matrix formalism is now a better tool to use in the discussion. For a mixed state that is probed in region A, the density matrix can be expressed as

$$\rho_A = \mathrm{tr}_B\ \rho_{AB} = \sum_n \alpha_n^2 |\psi_n^A\rangle\langle\psi_n^A|, \tag{3}$$

where the symbol $\alpha_n^2 \equiv p_n$ denotes the probability of an n-parton state. The basis $|\psi_n^A\rangle$ in (Eq. (2)) with states having a fixed number of n partons does not have interference between states with different number of partons due to the fact that this sort of interference is absent in the parton model. In this Schmidt decomposition, there can be an infinite number of terms (the Schmidt rank) in the sum shown in (Eq. (2)). A Schmidt rank one state is then a pure product state that does not include entanglement.

In the case of a mixed state, the probabilities corresponding to the different states described above can be used to define the von Neumann entropy of the mixed state given by

$$S = -\sum_n p_n \ln p_n. \tag{4}$$

It is the entanglement between regions **A** and **B** defined above that gives rise to what is called entanglement entropy here, and is related to Shannon entropy in information theory shown in (Eq. (4)). Hence the entanglement entropy can be determined form the QCD evolution equations that are used to evaluate the probabilities p_n. After the hard scattering takes place in the collision, the mixed quantum state characterized by the entanglement entropy (Eq. (4)) undergoes the evolution towards the final asymptotic state of hadrons that are measured by the detectors. This final state is characterized by the Boltzmann entropy. Further discussions of the relationship between Boltzmann entropy and entanglement entropy can be found in [14].

Studies of quantum entanglement and thermalization in atomic and condensed matter physics were shown to depend upon the quench properties, and that there is evidence for quantum propogation and information propagation [16, 21–23]. It is instructive to compare this with a quench induced by a high energy collision. The quench associated with the latter [17, 18] leads to the following interpretation. A quench produces a highly excited state of a Hamiltonian $H = H_0 + V(t)$ from what was the ground state of an unperturbed Hamiltonian H_0 originally. Here $V(t)$ is the term induced by the inelastic collision. Gluon exchange in the strong interaction induces the inelastic interaction, so the term $V(t)$ is seen to represent an effect of the pulse of the color field. The onset of this pulse in a hard scattering with a hardness scale Q, by the uncertainty principle, is $\tau \sim 1/Q$ where τ is the proper time. Since this time is short on the QCD scale, $\tau \ll 1/\Lambda$, the quench creates a highly excited multi-particle state. A short pulse of (chromo)electric field produces particles that have a thermal-like exponential spectra. The thermal spectrum in this case can be attributed to the emergence of an event horizon formed due to the acceleration induced by the electric field. Associated with this system is an effective temperature of $T \simeq (2\pi\tau)^{-1} \simeq Q/(2\pi)$ [24–27].

As shown in [17, 18] for a rapid quench (such as the one that occurs in a high-energy collision) in a $(1+1)$ dimensional CFT the entanglement entropy of a segment of length L first grows linearly in time, until $t \simeq L/2$, and then saturates at the value

$$S(t) \simeq \frac{c}{3} \ln \tau_0 + \frac{\pi c L}{12 \tau_0}. \tag{5}$$

where c is the conformal charge of the CFT, and τ_0^{-1} is the energy cutoff for the ultraviolet modes [14]. A sketch of the picture of the resulting thermalization from entanglement caused by the quench is shown in **Figure 2**.

The interpretation of the result (5) is the following [17, 18]. The quench leads to the production of entangled (quasi)particle pairs, since what used to be the ground state of the undisturbed Hamiltonian H_0 is a highly excited state of the Hamiltonian after the quench, $H = H_0 + V(t)$. The entangled pairs produced by the quench propagate along the light cone, and contribute to the entanglement entropy of the segment of length L if only one particle of the pair is detected within this segment. Shortly after the quench, only particle pairs produced near the boundary of the segment thus contribute to the entanglement, and the entanglement entropy is not extensive in the length L. However, at times $t > L/2$, even in the center of the segment one can detect a particle whose entangled partner is outside of the segment – this means that the entanglement entropy receives contributions from the entire segment, and should scale extensively in L in accord with the result (5). This scaling is a necessary condition for an effective thermalization.

For a quench induced by a high-energy collision, we sketch the resulting picture of thermalization from entanglement in **Figure 2**. Note that the hardest quasiparticle modes that propagate along the light cone thermalize first. For the softer

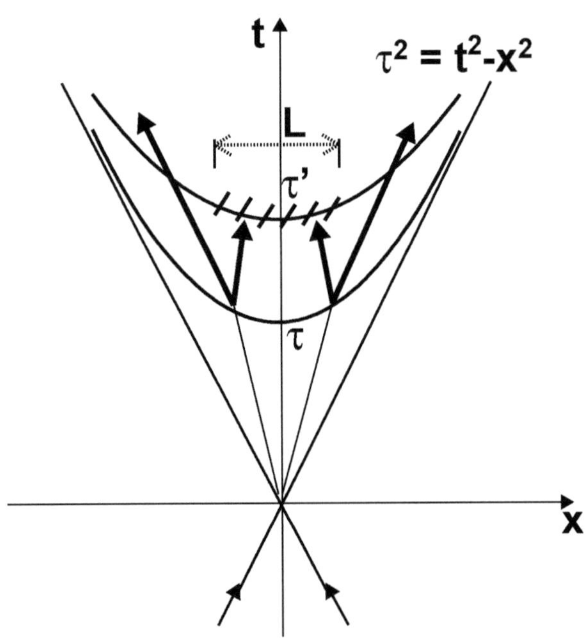

Figure 2.
An illustration of the onset of quantum thermalization through entanglement in a high energy pp collision. Time runs along the vertical axis, while space runs along the horizontal axis. The outermost lines define the light cone. The variables used are defined in the text. Entangled particle pairs that are produced at a proper time $\tau < \tau'$ contribute to the entanglement entropy in the interval of length L shown by the hashed segment of the curve. Figure from [14].

particles that propagate in the interior of the light cone, it takes a longer time to thermalize, that is, to exhibit an extensive scaling of the entropy. The detection of particles is assumed to be performed within the interval of length L (see Eq. (5)), corresponding to a limited range in (pseudo)rapidity. While (Eq. (5)) has been obtained in the framework of CFT, the simple physical interpretation of this result makes its broader validity quite likely.

It is instructive to point out the difference in the mechanisms of thermalization expected at weak and strong coupling. At weak coupling, the "bottom-up" thermalization mechanism [28] also yields an effective temperature $T \sim Q_s$ in inelastic high energy collisions. However the thermalization in this picture begins from the soft, low-momentum modes that eventually draw the energy from the harder modes; the thermalization of the hard, high-momentum modes is thus expected to take a parametrically long time proportional to the inverse power of the (small) coupling constant [28]. On the other hand, in strongly coupled entangled systems the process of thermalization is fast and determined by the size of the system and the parameters of the quench; moreover, it starts from the *hardest* modes resolved in the process. In the dual holographic description of conformal field theory, this process is described by the formation of trapped surface near the Minkowski boundary that then falls into the AdS bulk, corresponding to the spreading of thermalization from hard to soft modes [29, 30]. A similar picture emerges from the analysis of entanglement entropy in an expanding string [31], where the entropy has been found to have a thermal form with an effective temperature $T \sim 1/\tau$ at early time τ.

2.2 Charged hadron transverse momentum distribution

The discussion presented in the previous sections provide motivation to compare with experimental results from inelastic collision events at high energies. It also gives the opportunity to explore the possible relation between effective temperature and the hard scale of the collision. Consider proton-proton collisions data recorded by the LHC ATLAS collaboration at $\sqrt{s} = 13$ TeV center of mass energy yield multiple charged particles in the final state [32]. The data presented here corresponds to 151 μb^{-1} of integrated luminosity for charged particles with greater than 100 MeV/c transverse momenta and absolute pseudorapidity of less than 2.5. Events with two or more final state charged particles were selected in the analysis. Final state hadrons that originate in the primary pp interaction and that have a lifetime of greater than 30 ps were excluded from the final selected events in order to remove the presence of charged particles that have strangeness or are from heavier flavors.

The normalized charged hadron transverse momentum distribution is shown in **Figure 3**. The thermal component is shown by the exponential, red dashed curve; we parameterize it as

$$\frac{1}{p_T}\frac{d^2N_{ev}}{dp_T} = \mathrm{A}_{therm}\exp\left(-\mathrm{m_T}/\mathrm{T_{th}}\right), \tag{6}$$

where m_T, the hadron transverse mass, is defined as by $m_T \equiv \sqrt{m^2 + p_T^2}$ (m is the hadron mass; dominated by pions it is assumed), and T_{th} is an effective temperature. The hard scattering (power law, green solid curve) component is parameterized similar to [13],

$$\frac{1}{p_T}\frac{d^2N_{ev}}{dp_T} = \frac{\mathrm{A}_{hard}}{\left(1 + \frac{\mathrm{m_T^2}}{\mathrm{T^2 \cdot n}}\right)^{\mathrm{n}}}, \tag{7}$$

where T and n are parameters to be determined from the fit. The sum of the thermal and hard scattering contribution terms is shown by the blue solid curve in **Figure 3**.

The extracted value of the thermal temperature, $T_{th} = 0.17$ GeV describes well the experimental transverse momentum distribution, and it agrees with the temperature expected from the extrapolation of the relation [13] deduced at lower energies;

$$T_{th} = 0.098 \cdot \left(\sqrt{s/s_0}\right)^{0.06} \text{ GeV} \tag{8}$$

to the LHC 13 TeV collision energy; here $s_0 = 1\ \text{GeV}^2$. Similarly, the hard scale temperature parameter T is [13]

$$T = 0.409 \cdot \left(\sqrt{s/s_0}\right)^{0.06} \text{ GeV}. \tag{9}$$

It's interesting that the parameterizations (Eqs. (8) and (9)) imply that the effective thermal temperature T_{th} is proportional to the hard scale temperature parameter T, which is in agreement with the Section 2.1 discussion.

The fits to the charged hadron transverse momentum distribution in **Figure 3** yields the hard scale temperature parameter $T = 0.72$ GeV and $n = 3.1$, in agreement with the extrapolation of (Eq. (9)) to 13 TeV pp collision energy, but with a smaller value of n. This reflecting the slower fall-off of the transverse momentum distribution at the LHC energy.

The integral of the area under the fit curves carries important information about entanglement in these and other in high energy physics processes. Defining the ratio R of the integral under the power law (hard scattering component) curve, I_p and the sum of the integrals of the exponential (thermal component) curve, I_e and power law curve of the fit in **Figure 3**:

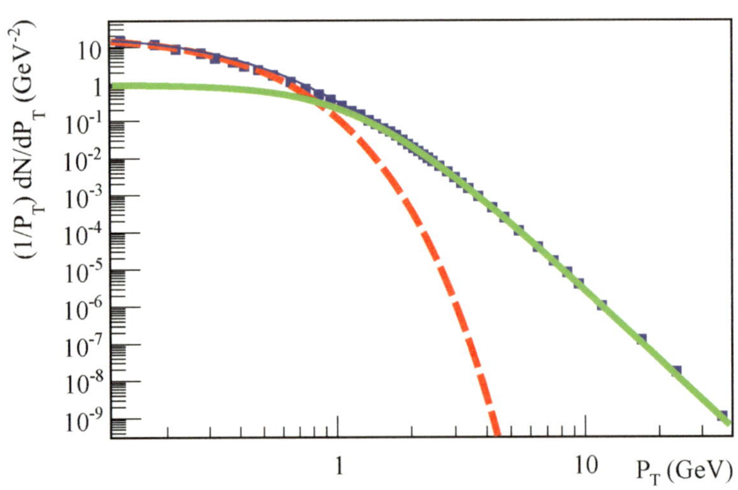

Figure 3.
Normalized transverse momentum distribution of charged hadrons from $\sqrt{s} = 13$ TeV pp collisions. The curves shown are exponential (red dashed) representing the thermal component of the distribution, and power law (green solid) corresponding to the hard scattering contribution. The superposition of these two contributions are also shown (blue, thin solid). Figure from [14], and data is from [32].

$$R = \frac{\text{power}}{\text{power} + \text{exponential}} = \frac{I_p}{I_p + I_e} \tag{10}$$

The calculation yields the value of $R \simeq 0.16 \pm 0.05$, in agreement (within the uncertainty interval) with the ratio calculated from the charged hadron spectra in inelastic proton-proton collisions at ISR energies of $\sqrt{s} = 23, 31, 45,$ and 53 GeV [20] even given the large beam energy difference between the LHC and the ISR accelerators.

2.3 Diffractive events and di-muon pair transverse momentum distribution in proton-proton collisions

Diffractive proton-proton (pp) collision events at the LHC can proceed through the photon-photon ($\gamma\gamma$) interactions shown in (Eq. (11)). Both X' and X'' can be final state protons from the collision, or the products X', X'' of their diffractive dissociation (single diffraction (in which one of the incident protons dissociates into an inelastic state), and double diffraction (in which both of the incident protons dissociate)). Measurements from the ATLAS collaboration [33] of the reaction

$$pp(\gamma\gamma) \rightarrow \mu^+\mu^- X'X'' \tag{11}$$

at $\sqrt{s} = 13$ TeV center of mass energy in pp collisions are studied. Selection of the exclusive $\gamma\gamma \rightarrow \mu^+\mu^-$ process was implemented by only including events that have both muon tracks (μ^+ and μ^-) while at the same time excluding events that show additional charged particle activity in the central region of the detector. Transverse momenta of greater than 400 MeV were used in the ATLAS analysis, with pseudorapidity range the same as that of the charged hadron analysis described in subSection 2.2. In the most recent ATLAS analysis of the reaction (11) only diffractive events that proceed through the $\gamma\gamma$ scattering were selected [33].

Figure 4 shows the transverse momentum distribution in the case of $\gamma\gamma$ production of di-muon pairs in pp collisions at $\sqrt{s} = 13$ TeV center of mass energy. As can

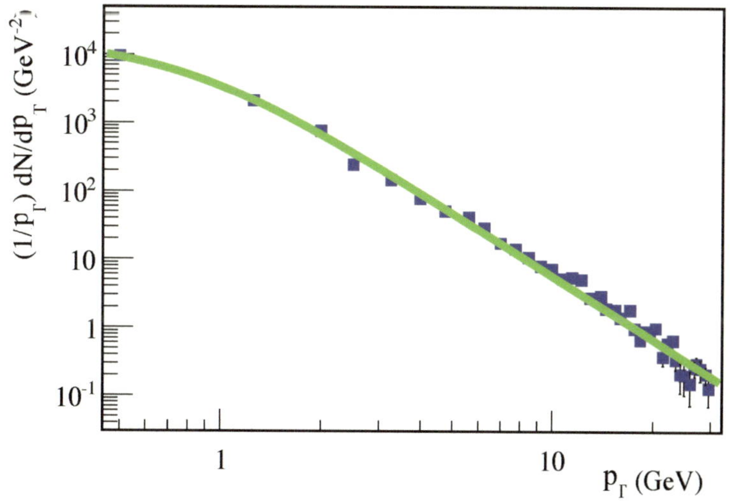

Figure 4.
The (normalized) transverse momentum diffractive scattering event distribution $\frac{1}{p_T}\frac{dN_{\mu\mu}}{dp_T}$ *in units of* GeV^{-2} *for the reaction of (Eq. (11)) showing the absence of a thermal component to the distribution. The curve shown (green, solid) is the power law contribution corresponding to the hard-scattering process. Data from [33], figure from [14].*

be seen there, the hard scattering term alone describes well the distribution, and the thermal (exponential) component is absent. As discussed already in this chapter and in [13–15], diffractive events are expected to have a suppressed thermal (exponential) component due to the fact that in these diffractive processes the photon interacts coherently with the entire proton, and no entanglement entropy is expected. This was discussed in Section 2.1. As the presence of the thermal component in this approach is the consequence of the entanglement, we expect it to be absent in diffractive events, as confirmed in **Figure 4**. Furthermore, the ratio R defined in the previous section in this case is $R \simeq 1$, in agreement with the theoretical expectations and the previous data for $\gamma\gamma$ scattering at OPAL at $\sqrt{s} = 15$ and 35 GeV that also show no thermal component. R is then equal to one within experimental uncertainty.

2.4 Combined Higgs boson decays to $\gamma\gamma$, $ZZ^* \rightarrow 4l$, and $b\bar{b}$

The Higgs boson differential transverse momentum cross section is undoubtedly adequately described by perturbation theory (see [34] for a review). An investigation is undertaken to determine whether the thermalization process due to entanglement is present in this system. The Higgs boson differential cross sections (differential in transverse momentum p_T) have been measured by both ATLAS and CMS collaborations [35–37] and most recently from [38].

In **Figure 5** the transverse momentum distribution of the Higgs bosons is shown in the range from 5 GeV to 700 GeV for combined ATLAS and CMS data at 13 TeV pp collision energy. As can be seen from **Figure 5**, there clearly are both the hard scattering (power law) and thermal (exponential) components in the transverse momentum distribution, similarly to the case explored in Section 2.2. Not surprisingly, the separation between the hard and thermal components is even more defined due to the much larger range of the available transverse momenta.

Interestingly, the ratio R defined by (Eq. (10)) and extracted from **Figure 5** is $R = 0.15 \pm 0.03$ that is very close to the one determined from the charged hadron distribution in proton-proton collisions studied in Section 2.2, $R = 0.16 \pm 0.05$.

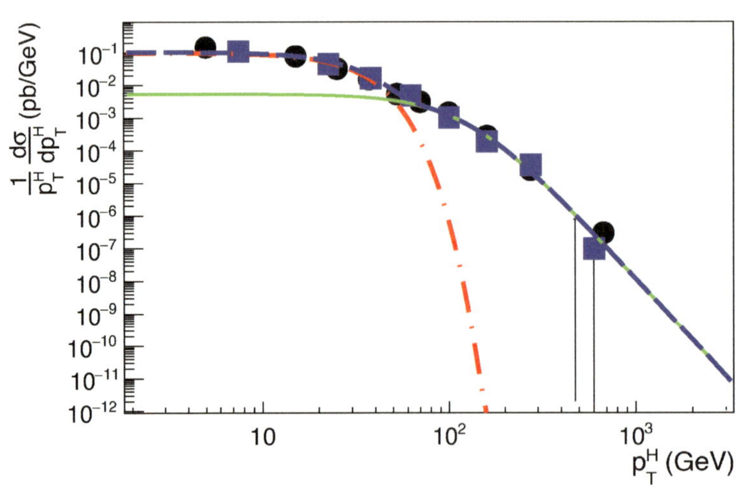

Figure 5.
Normalized fiducial Higgs differential cross section versus transverse momentum reconstructed from the combination of $H \rightarrow \gamma\gamma$, four leptons, and bbarb decay in proton-proton collisions at $\sqrt{s} = 13$ TeV from both the ATLAS and CMS collaborations [38].

2.5 Discussion: entanglement entropy in proton-proton collisions

The material presented in Section 2 provide evidence for an unconventional mechanism of apparent thermalization in high energy pp collisions. The data shows that the effective thermal temperature T_{th} is non-universal and that it is proportional to the hard scale temperature parameter of the collision T, that is, to the momentum transfer, with $T \simeq 4.2 \ T_{th}$. Strikingly, this conclusion seems to apply even to the Higgs boson production, suggesting that even in this very hard process the QCD radiation may be affected by thermalization. Moreover, we have found that the thermal component of the spectrum is entirely absent in diffractive production (even though many hadrons are still produced in this case) – this again points to the non-universal, process-dependent, nature of thermalization.

The theory and the analyses of the data discussed in Section 2 appear to be consistent with the proposition that thermalization in these high energy collisions is induced by quantum entanglement. That the effective temperature determined from the data is proportional to the momentum transfer Q in the collision that provides the UV cutoff for the quantum modes, as expected. Notably, inclusive charged hadron and Higgs boson transverse momentum distributions, in which the typical momentum transfers are vastly different are in agreement in this analysis. It is seen that the thermal component is present in both cases, event though the values of the effective temperature differ by over an order of magnitude.[1]

In diffractive events studied in Section 2, it is clearly seen that where studies of the coherent response of the entire proton in this scattering, there is no associated entanglement entropy [15], and that therefore there should be no thermal component to the transverse momentum distribution. The data confirms this prediction in diffractive Drell-Yan production analyzed in this section, as well as by the diffractive deep-inelastic scattering data shown in [20].

The findings presented here appear to support the proposition that a deep connection between quantum entanglement and thermalization in high-energy hadron collisions, and that this proposed link should be further investigated. Possibilities include the following as non-exhaustive examples. Combining measurements of the structure functions with the study of hadronic final states, especially in the target fragmentation region in deep inelastic scattering at the future Electron Ion Collider. Studies of the thermal component and the corresponding effective temperature in hard processes characterized by different momentum transfers in proton-proton, proton-nucleus and nucleus-nucleus collisions at RHIC and the LHC. Already, analysis of Pb–Pb HI collision data also points to a picture of thermalization as a result of quantum entanglement at high energies [9]. An investigation of the dependence of the apparent thermalization on rapidity – as depicted in **Figure 2**, suggesting that the thermal component and the corresponding effective temperature in hard processes characterized by different momentum transfer would be interesting. It suggests that thermalization is achieved faster if a measurement is performed in a smaller rapidity interval.

3. Entanglement entropy and thermal behavior in the electroweak interaction

The material and discussion in Section 2 supporting a picture of thermalization in hadronic physics due to quantum entanglement motivates an investigation of

---

[1] It is once again emphasized that this does not imply that the Higgs boson is produced thermally, but rather that its transverse momentum distribution is affected by thermal radiation due to entanglement.

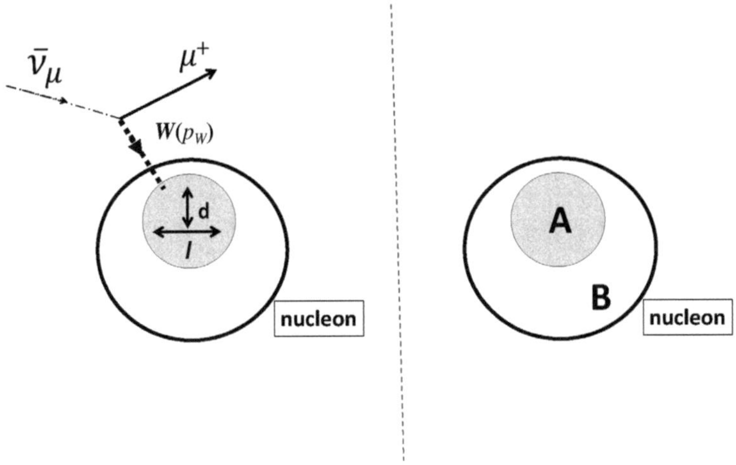

Figure 6.
(Left side) The depiction of antineutrino scattering from a nucleon via emission of a W boson with an exiting muon in the final state. The W boson samples a partial region of the nucleon, not the entire nucleon, as explained in the text. (Right side) The region of the nucleon sampled by the interacting W boson is denoted as region A. The nucleon spectator region not probed by the boson is region B. Figure from [5].

whether this same connection is manifested in weak interactions mediated by massive vector bosons. In this section that study, taken mainly from [5], is made using charged-current weak interaction processes such as

$$\bar{\nu}_\mu + N \to \mu^+ + \pi^0 + X \tag{12}$$

Similar to the partial probing of the nucleon wave function described in Section 2 the vector boson in this investigation probes only a part of the nucleon wave function, again denoted by the region **A** in **Figure 6**. This probed region has a transverse size of approximately $d = h/p_W$, and a longitudinal size of approximately $l = (mx)^{-1}$ [1, 2, 14]. In this analysis, h is Planck's constant, p_W is the boson's momentum, x is the momentum fraction carried by the struck quark in the interaction (Bjorken-x), and m is the nucleon mass. Within the struck nucleon, the probed region **A** is complementary to the spectator region **B** that is not probed in the interaction. The entire space within the nucleon (a pure state) is then **A**∪**B**. In this present analysis, as in [14], thermal behavior is attributed to the quantum entanglement between regions **A** and **B** as depicted in in **Figure 6**.

In this current analysis, we test the hypothesis, albeit disfavored by the conventional mechanism of thermalization, that the thermal feature found in the low-p_T region (corresponding to measurement at late times) of the momentum distribution can instead be attributed to the sub-nucleonic entanglement induced by collisions at high energies. This is the gist of the study using charged-current anti-neutrino interactions at the intensity frontier in particle physics. The claim from the the first two sections of this chapter is further strengthened by the demonstration that when the nucleus as a whole is scattered by the W boson so that no sub-nucleonic entanglement is produced, the thermal feature is absent from the spectrum, as expected. And that when quantum entanglement exists in the process, thermalization is present in the momentum distribution.

3.1 Charged current weak interactions: analysis and results

We begin by considering neutral pion production in charged-current antineutrino interactions with a CH (hydrocarbon scintillator) target; see (Eq. (12)). This experimental data includes the total inclusive charged current weak interaction differential cross sections [39, 40] measurements at $1.5\ GeV < E_\nu < 10\ GeV$ [39] and data at $E_\nu = 3.6\ GeV$ [40]. The analysis results from both references, and from [5], are described in this present study. A conversion from pion kinetic energy (T_π) published in [39] to pion momentum published in [40] is made using the expression

$$\frac{d\sigma}{dp_\pi} = \frac{p_\pi c^2}{T_\pi + m_{0,\pi}c^2}\frac{d\sigma}{dT_\pi}. \tag{13}$$

The relativistic kinetic energy is related to the pion rest mass, $m_{0,\pi}c^2$, by

$$T_\pi = (\gamma - 1)m_{0,\pi}c^2 \tag{14}$$

where $\gamma = 1/\sqrt{1 - v^2/c^2}$, with v the pion velocity in this case. We will compare the above results against the inclusive charged-current coherent pion production differential cross sections given in [41].

The normalized differential cross section that is used to describe the thermal behavior from the interaction is given by a very similar formula as in subSection 2.2 but here using

$$\frac{1}{p_\pi}\frac{d\sigma}{p_\pi} = A_{\text{thermal}}e^{(-E_\pi/T_{\text{thermal}})} \tag{15}$$

where p_π ($E_\pi = \sqrt{m_\pi^2 + p_\pi^2}$) is the pion momentum (energy) and where the Mandelstam variable s is approximately equal to $m^2 + 2E_\nu m$ The hard-scattering part of the normalized momentum distribution is given by

$$\frac{1}{p_\pi}\frac{d\sigma}{p_\pi} = A_{\text{hard}}\left(1 + \frac{m_\pi^2}{T_{\text{hard}}^2 \cdot n}\right)^{-n} \tag{16}$$

where n a power law scaling parameter. These equations are also discussed in [14, 42].

The CERN ROOT fitting program is used to fit these expressions to the MINERvA results. A total of five parameters are used in the fitting procedure: $T_{\text{thermal}}, T_{\text{hard}}, n, A_{\text{hard}}$, and A_{thermal}. In each case, the reduced chi-squared statistic and the fitting parameters with their associated uncertainties are recorded.

The results of fitting the thermal and hard scattering components to the distribution in the analysis using data from the MINERνA collaboration [39, 40] are shown in **Figure 7**. As can seen from the fit, there are separate thermal (red-dashed) and hard-scattering (green-full) components in the full momentum distributions. The solid blue curve is the superposition of the exponential and power law fits.

Final state interactions (FSI) are modeled using the GENIE Monte Carlo program [43] in the anayses described in [39, 40]. They show that the larger FSI effects on the data are at low pion momenta. These effects are small compared with the statistical and other systematic uncertainties from the analysis, and did not affect the fits and conclusions drawn in this present study.

Now consider the resulting momentum distribution when the process of antineutrino scattering is from the entire nucleus, and not from a partial region of the nucleon as described above. That is, when the antineutrino scatters from the nucleus coherently, as in

$$\overline{\nu}_\mu + A \rightarrow \mu^+ + \pi^- + A. \quad (17)$$

In this charged current weak interaction, there is no entanglement between different parts of a struck nucleon, and no thermal component to the momentum distribution of the single produced pion is expected. It is this description of the interaction that is supported by the coherent scattering data from the MINERνA collaboration [41], as shown in **Figure 8**. Only the hard scattering (power law) fit component is needed to describe the momentum distribution. The absence of a thermal (exponential) fit component is due to the absence of entanglement in the proposition presented in this present work.

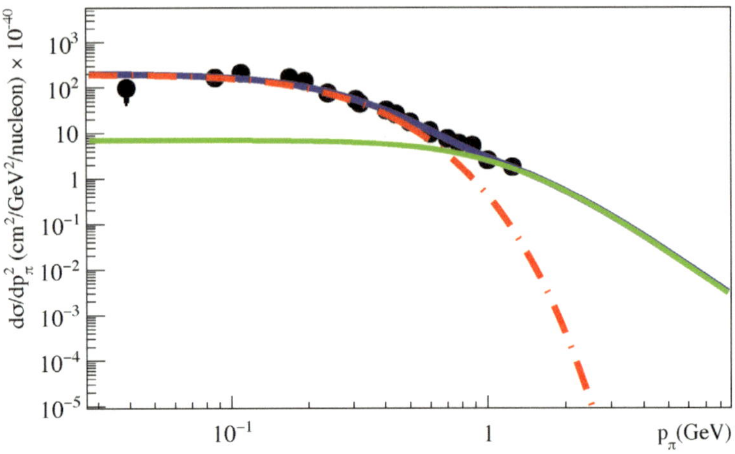

Figure 7.
Antineutrino differential cross section for scattering against ahydrocarbon nuclei with resulting charged current pion production. The dashed (red) line fit to the data is the thermal component fit and the thick solid (green) line shows the hard component fit. The combined thermal and hard scattering thin solid (blue) line best fits to the data. Data taken from [39, 40]. Plot taken from [5].

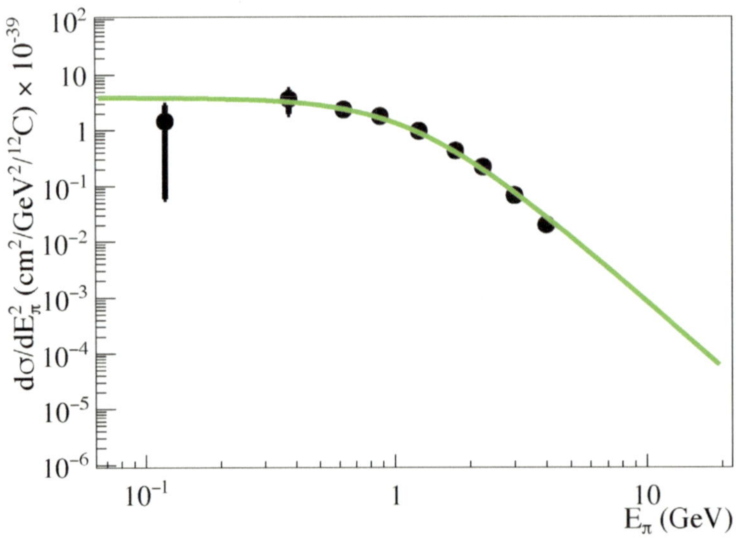

Figure 8.
Coherent scattering of the antineutrino from the hydrocarbon scintillator nuclei results in the momentum distribution shown here. The differential cross section is well described by a hard-scattering component (solid green line) alone, as expected in the absence of entanglement. The data is from [41]. The figure is from [5].

R	Process	Reference
0.16 ± 0.05	$pp \to$ charged hadrons	[14, 44]
0.15 ± 0.05	$pp \to \mathrm{H} \to \gamma\gamma$	[14, 44]
0.23 ± 0.05	$pp \to \mathrm{H} \to 4l(e,\mu)$	[14, 44]
1.00 ± 0.02	$pp(\gamma\gamma) \to (\mu\mu)\,\mathrm{X'X''}$	[14, 44]
0.13 ± 0.03	$\overline{\nu}_\mu + N \to \mu^+ + \pi^0 + X$	[5]
1.00 ± 0.05	$\overline{\nu}_\mu + ^{12C} \to \mu^+ + \pi^- + ^{12}C$	[5]

Table 1.
The ratio R is defined in (Eq. (10)) for different processes as shown. The results listed indicate that the thermal behavior due to entanglement entropy is independent of the interaction (strong or electroweak) but process dependent.

4. Conclusion

R (Eq. (10)) is computed from the integral of the combined fit, which combines the hard-scattering function (Eq. (16)) and the exponential function (Eq. (15)). The *R* values obtained in charged-current weak interactions are consistent with values obtained for *pp* collisions [14]. And as stated in Section 2 they are also in agreement (within experimental uncertainly) with values obtained from low energy ISR and HERA data [20]. **Table 1** presents a compilation of the ratio R (defined by (10) for the processes considered in this present study.

The results presented in this study support those given in [1, 2, 14, 18], namely that quantum entanglement in hadrons is what gives rise to the thermal behavior observed in hadronic collisions and, as the new results from charged-current neu trino scattering presented here suggest, that the thermalization process from entanglement, while process dependent, is interaction independent.

Author details

Oliver Keith Baker
Department of Physics, Yale University, New Haven, CT, USA

*Address all correspondence to: oliver.baker@yale.edu

References

[1] D. E. Kharzeev, E. M. Levin, Deep inelastic scattering as a probe of entanglement, Phys. Rev. D 95 (11) (2017) 114008.

[2] A. A. Bylinkin, D. E. Kharzeev, A. A. Rostovtsev, The origin of thermal component in the transverse momentum spectra in high energy hadronic processes, Int. J. Mod. Phys. E 23 (12) (2014) 1450083.

[3] , R. Hagedorn, Statistical thermodynamics of strong interactions at high energies, Nuovo Cim. Suppl. 3, 147 (1965); Hadronic matter near the boiling point, Nuovo Cim. A, 56, 1027 (1965).

[4] , F. Becattini, An introduction to the Statistical Hadronization Model, arXiv: 0901.3643 [hep-ph] (2009).

[5] G. Iskander, J. Pan, M. Tyler, C. Weber, O.K. Baker, Quantum Entanglement and Thermal Behavior in Charged-Current Weak Interactions, Phys Lett **B 811**, 135948 (2020).

[6] G. Ramos, M. Machado, Investigating entanglement entropy at small-x in dis off protons and nuclei, Phys. Rev. D 101 (2020) 074040.

[7] M. Mezei, W. van der Schee, Black holes often saturate entanglement entropy the fastest, Phys. Rev. Lett. 124 (2020) 201601–201606.

[8] T. T. Takahashi, Y. Kanada-En'yo, Lattice QCD study of static quark and antiquark correlations via entanglement entropies, Phys. Rev. D 100 (2019) 114502.

[9] X. Feal, C. Pajares, R. A. Vazquez, Thermal behavior and entanglement in Pb-Pb and pp collisions, Phys. Rev. C 99 (1) (2019) 015205.

[10] C. M. Ho, S. D. H. Hsu, Entanglement and fast quantum thermalization in heavy ion collisions, Mod. Phys. Lett. A 31 (18) (2016) 1650110.

[11] P. Castorina, A. Iorio, D. Lanteri, P. Lukes, Gluon shadowing and nuclear entanglement, arXiv:2003.00112v1 [hep-ph] (Feb. 2020).

[12] S. Beane, P. Ehlers, Chiral symmetry breaking, entanglement, and the nucleon spin decomposition, Report Number NT@UW-19-06; arXiv: 1905.03295v1 [hep-ph] (May 2019).

[13] A.A. Bylinkin, D.E. Kharzeev, and A.A. Rostovtsev, "The origin of thermal component in the transverse momentum spectra in high energy hadronic processes", Physics Preprint Archives arXiv:1407.4087, 1 (2014).

[14] O. K. Baker, D. E. Kharzeev, Thermal radiation and entanglement in proton-proton collisions at the LHC, Phys. Rev. D 98 (5) (2018) 054007.

[15] , D.E. Kharzeev, and E.M.Levin, "Deep inelastic scattering as a probe of entanglement" Phys Rev **D 95**, Number 11, 114008 (2017).

[16] A.M. Kaufman, M.E. Tai, A. Lukin, M. Rispoli, R. Schittko, P.M. Preiss, M. Greiner, "Quantum thermalization through entanglement in an isolated many-body system", Science **353**, Number 6301, 794-800 (2016).

[17] P. Calabrese and J. Cardy, "Evolution of entanglement entropy in one-dimensional systems", Journal of Statistical Mechanics: Theory and Experiment **2005**, Number 04, P04010 (2005).

[18] P. Calabrese and J. Cardy, "Quantum quenches in 1+1 dimensional conformal

field theories", Journal of Statistical Mechanics: Theory and Experiment **2016**, Number 06, 064003 (2016).

[19] P. Calabrese and J. Cardy, "Quantum quenches in extended systems", Journal of Statistical Mechanics: Theory and Experiment **2007**, Number 06, P06008 (2007).

[20] A.A., Bylinkin and A.A. Rostovtsev, "Role of quarks in hadroproduction in high energy collisions", Physics Preprint Archives arXiv:1404.7302, 1 (2014).

[21] P. Jurcevic et al. Nature 511.7508, 202 (2014).

[22] M. Rigol, V. Dunjko, V. Yurovsky, and M. Olshanii Phys. Rev. Lett., 98(5), 050405 (2007).

[23] Z. X. Gong, L.M. Duan, New Journal of Physics, 15(11), 113051 (2013).

[24] D.E Kharzeev and K. Tuchin, "From color glass condensate to quark–gluon plasma through the event horizon", Nucl. Phys. **A753** Number 3-4, 316 (2005).

[25] G. Dunne and T. Hall, "QED effective action in time dependent electric backgrounds", Phys. Rev. **D58**, Number 10, 105022 (1998).

[26] P. Castorina, D. Kharzeev and H. Satz, Eur. Phys. J. C **52**, 187 (2007).

[27] D.E. Kharzeev, E. Levin, and K. Tuchin, "Multiparticle production and thermalization in high-energy QCD", Phys. Rev. **C75**, Number 4, 044903 (2007).

[28] R. Baier, A. H. Mueller, D. Schiff and D. T. Son, Phys. Lett. B **502**, 51 (2001) doi:10.1016/S0370-2693(01)00191-5 [hep-ph/0009237].

[29] S. Lin and E. Shuryak, Phys. Rev. D **77**, 085013 (2008) doi:10.1103/PhysRevD.77.085013 [hep-ph/0610168].

[30] V. Balasubramanian *et al.*, Phys. Rev. D **84**, 026010 (2011) doi:10.1103/PhysRevD.84.026010 [arXiv:1103.2683 [hep-th] (2011)].

[31] J. Berges, S. Floerchinger and R. Venugopalan, arXiv:1707.05338 [hep-ph] (2017).

[32] The ATLAS collaboration, "Charged-particle distributions at low transverse momentum in $\sqrt{s} = 13$ TeV pp interactions measured with the ATLAS detector at the LHC", Eur. Phys. J. **C76**, 502 (2016); The ATLAS collaboration, Charged-particle distributions in $\sqrt{s} = 13$ TeV pp interactions measured with the ATLAS detector at the LHC', Phys Lett. **B758**, 67 (2016).

[33] The ATLAS collaboration, Measurement of the exclusive $\gamma\gamma \rightarrow \mu\mu$ process in proton-proton collisions at $\sqrt{s} = 13$ TeV with the ATLAS detector, Phys. Lett. B777, 303 (2018); arXiv: 1708.04053 (2017).

[34] S. Dittmaier *et al.* [LHC Higgs Cross Section Working Group], doi:10.5170/CERN-2011-002 arXiv:1101.0593 [hep-ph] (2011).

[35] The ATLAS collaboration, "Measurement of inclusive and differential cross sections in the $H \rightarrow 4l$ decay channel in pp collisions at $\sqrt{s} = 13$ TeV with the ATLAS detector", J. High Energ. Phys., 2017: 132 (2017).

[36] Eleni Mountricha on behalf of the ATLAS collaboration, "Higgs Measurements in High Resolution Channels with ATLAS", LHCP2017, Large Hadron Collider Physics Conference, Shanghai, China, June 2017 (2017).

[37] J. Tao on behalf of the CMS collaboration, "Higgs boson measurements in high resolution channels with CMS", Physics Preprint Archives arXiv:1708.09215, 1 (2017).

[38] ATLAS and CMS collaborations, European Physical Society Conference on High Energy Physics (EPS-HEP) , July 10-17 (2019), https://indico.cern.ch/event/577856/ (2019).

[39] C. L. McGivern, et al., "Cross sections for neutrino and antineutrino induced pion production on hydrocarbon in the few-GeV region using MINERvA," Phys. Rev. D 94, 052005 (2016).

[40] T. Le, et al., "Single neutral pion production by charged-current ν_μ interactions on hydrocarbon at $\langle E_\nu \rangle =$ 3.6 GeV", Phys. Lett. B 749, 130 (2015).

[41] A. Mislivec, et al., "Measurement of total and differential cross sections of neutrino and antineutrino coherent π production on carbon", Phys. Rev. D 97, 032014 (2018).

[42] A. Bylinkin, A. Rostovtsev, "Role of quarks in hadroproduction in high energy collisions", Nucl. Phys. B 888, 65 (2014).

[43] C. Andreopoulos, et al., "The GENIE Neutrino Monte Carlo Generator", Nucl. Instrum. Meth. 614 (2010) 87–104.

[44] C. Weber, O. K. Baker, D. Kharzeev, "Quantum Entanglement and Proton- Proton Collisions at the LHC", SUSY International Conference, Corpus Christie, TX (2019) (Jul. 2019).